Northwestern Handbook of Surgical Procedures

Richard H. Bell, Jr., M.D.
*Department of Surgery, Feinberg School of Medicine,
Northwestern University, Chicago, Ilinois, U.S.A.*

Dixon B. Kaufman, M.D., Ph.D.
*Department of Surgery, Feinberg School of Medicine,
Northwestern University, Chicago, Ilinois, U.S.A.*

Illustrations by Simon Kimm, M.D.

LANDES
BIOSCIENCE
GEORGETOWN, TEXAS
U.S.A.

VADEMECUM
Northwestern Handbook of Surgical Procedures
LANDES BIOSCIENCE
Georgetown, Texas U.S.A.

Please address all inquiries to the Publisher:
Landes Bioscience, 810 S. Church Street, Georgetown, Texas, U.S.A. 78626
Phone: 512/ 863 7762; FAX: 512/ 863 0081

ISBN: 1-57059-684-0

Library of Congress Cataloging-in-Publication Data

Bell, Richard H., 1946-
    Northwestern handbook of surgical procedures / Richard H. Bell Jr., Dixon B. Kaufman ; illustrations by Simon Kimm.
        p. ; cm. -- (Vademecum)
    ISBN 1-57059-684-0
    1. Surgery, Operative--Handbooks, manuals, etc. I. Kaufman, Dixon B. II. Northwestern University (Evanston, Ill.) III. Title. IV. Title: Handbook of surgical procedures. V. Series.
        [DNLM: 1. Surgical Procedures, Operative--methods--Handbooks.   WO 39 B435n 2005]
    RD37.B45 2005
    617'.91--dc22
                                    2005012436

Dedication

To today's students of surgery, whose skills will mature to the benefit of man.

Contents

Preface xii

Section 1: Gastrointestinal 1

1. Esophageal Diverticulectomy: Zenker's 2
2. Heller Myotomy: Laparoscopic 4
3. Thoracic Esophageal Perforation Repair 7
4. Antireflux Procedure: Laparoscopic (Nissen) 9
5. Gastrostomy: Open 11
6. Gastrectomy: Total 14
7. Gastrectomy: Subtotal or Partial 18
8. Gastric Bypass: Roux-en-Y 22
9. Truncal Vagotomy and Pyloroplasty 24
10. Highly Selective (Parietal Cell) Vagotomy 26
11. Perforated Duodenal Ulcer Repair: Omental Patch 28
12. Major Hepatic Laceration: Open Repair 29
13. Hepaticojejunostomy: Roux-en-Y 34
14. Cholecystectomy with Cholangiography: Open 38
15. Cholecystectomy with Cholangiogram: Laparoscopic 41
16. Common Bile Duct Exploration: Open 45
17. Common Bile Duct Exploration: Laparoscopic 47
18. Transduodenal Sphincteroplasty 50
19. Pancreatic Necrosis: Debridement 53
20. Pancreaticoduodenectomy: Whipple Procedure 55
21. Distal Pancreatectomy and Splenectomy 60
22. Pancreatic Cystogastrostomy 62
23. Longitudinal Pancreaticojejunostomy: Puestow Procedure 64
24. Duodenum-Preserving Subtotal Pancreatic Head Resection:
    Frey Procedure 67
25. Splenectomy: Open 71
26. Splenectomy: Laparoscopic 73
27. Splenorrhaphy: Open 76
28. Small Bowel Resection and Anastomosis (Enterectomy): Open 79
29. Enterolysis for Small Bowel Obstruction: Open 82
30. Appendectomy: Open 84
31. Appendectomy: Laparoscopic 86
32. Ileostomy: Open Loop 88
33. Hemicolectomy (Right): Open 90
34. Hemicolectomy (Right): Laparoscopic 93
35. Colostomy Closure 96
36. Colostomy: End Sigmoid with Hartmann's Pouch 98
37. Colostomy: Transverse Loop 101

38. Sigmoid Colectomy: Open 104
39. Proctocolectomy with Ileal Pouch: Anal Anastomosis 106
40. Proctocolectomy: Total with Ileostomy 110
41. Anal Fistulotomy 113
42. Anal Fissure: Lateral Internal Sphincterotomy 116
43. Anorectal Abscess: Drainage Procedure 118
44. Internal Hemorrhoids: Band Ligation 120
45. Inguinal Hernia Repair with Mesh: Open 122
46. Inguinal Hernia Laparoscopic Repair: Extraperitoneal Approach ... 124
47. Ventral Hernia Repair: Open 128
48. Ventral Hernia Repair: Laparoscopic 130
49. Exploratory Laparotomy: Open 133

Section 2: Endocrine 137
50. Adrenalectomy: Laparoscopic 138
51. Pancreatic Endocrine Tumor Enucleation 141
52. Parathyroid Adenoma Excision 144
53. Radioguided Parathyroidectomy: Minimally Invasive 147
54. Thyroid Lobectomy and Total Thyroidectomy 149
55. Modified Neck Dissection 153

Section 3: Surgical Oncology 157
56. Transanal Excision of Rectal Tumor 158
57. Abdominoperineal Resection 160
58. Right Hepatic Lobectomy 163
59. Axillary Lymphadenectomy 166
60. Inguinal Lymphadenectomy 169
61. Breast Biopsy after Needle Localization 172
62. Lymphatic Mapping and Sentinel Node Biopsy 174
63. Partial Mastectomy and Axillary Dissection 176
64. Modified Radical Mastectomy 179
65. Simple Mastectomy 182
66. Major Excision and Repair/Graft for Skin Neoplasms 184
67. Sentinel Lymph Node Biopsy for Melanoma 187
68. Radical Excision of Soft Tissue Tumor (Sarcoma) 190

Section 4: Plastic Surgery 193
69. Burn Debridement and/or Grafting 194
70. Split-Thickness Skin Grafts 196
71. Debride/Suture Major Peripheral Wounds 198
72. Repairing Minor Wounds 200
73. Removal of Moles and Small Skin Tumors 202
74. Removal of Subcutaneous Small Tumors, Cysts
    and Foreign Bodies 204

Section 5: Cardiothoracic Surgery 207

75. Esophagectomy: Ivor-Lewis 208
76. Esophagectomy: Left Transthoracic 215
77. Esophagectomy: Transhiatal 220
78. Mediastinoscopy: Cervical 226
79. Lung Biopsy: Thoracoscopic 229
80. Pulmonary Lobectomy: Open 233
81. Pneumonectomy 236
82. Pleurodesis: Thoracoscopic 239
83. Tracheostomy 241

Section 6: Transplantation 245

84. Arteriovenous Graft (AVG) 246
85. Primary Radial Artery-Cephalic Vein Fistula
    for Hemodialysis Access 249
86. Laparoscopic Donor Nephrectomy 252
87. Kidney Transplantation 255
88. Distal Splenorenal (Warren) Shunt 258
89. H-Interposition Mesocaval Shunt 261
90. Portacaval Shunts 264
91. Liver Transplantation 268
92. Pancreas Transplantation 274

Section 7: Vascular Surgery 277

93. Carotid Endarterectomy 278
94. Repair Infrarenal Aortic Aneurysm: Elective 281
95. Repair Infrarenal Aortic Aneurysm: Emergent for Rupture 284
96. Endovascular Repair of Infrarenal Aortic Aneurysm 287
97. Aortofemoral Bypass for Obstructive Disease 290
98. Axillofemoral Bypass 292
99. Femorofemoral Bypass 297
100. Femoral-Popliteal Bypass with a Vein or Prosthetic Graft 300
101. Composite Sequential Bypass 303
102. Infrapopliteal Bypass: Vein or Prosthetic 305
103. Lower Extremity Thrombectomy/Embolectomy 308
104. Repair Popliteal Aneurysm: Emergent (Thrombosed) 311
105. Exploration for Postoperative Thrombosis 313
106. Fasciotomy: Lower Extremity 315
107. Toe Amputation 317
108. Transmetatarsal Amputation 319
109. Below Knee Amputation (BKA) 321
110. Above Knee Amputation (AKA) 323
111. Varicose Veins 325

Editors

Richard H. Bell, Jr., M.D.
Gastrointestinal and Endocrine
Chapters 7, 10, 22-24

David Fullerton, M.D.
Cardiothoracic Surgery
Chapters 80, 82

Dixon B. Kaufman, M.D., Ph.D.
Transplantation
Chapters 87, 92

Thomas Mustoe, M.D.
Plastic Surgery
Chapter 72

William H. Pearce, M.D.
Vascular Surgery
Chapters 101, 111

Mark S. Talamonti, M.D.
Surgical Oncology
Chapters 13, 20

David P. Winchester, M.D.
Gastrointestinal

Department of Surgery, Feinberg School of Medicine,
Northwestern University, Chicago, Ilinois, U.S.A.

Illustrations by Simon Kimm, M.D.

Contributors

Michael Abecassis, M.D.
Chapters 88-90

Peter Angelos, M.D., Ph.D.
Chapters 50, 54, 55

Ermilo Barrera, Jr., M.D.
Chapters 11, 45

Richard S. Berk, M.D.
Chapters 5, 37

Kevin Bethke, M.D.
Chapters 59, 60

Malcolm M. Bilimoria, M.D.
Chapter 6, 25, 26

Christopher Bulger, M.D.
Chapter 100

Joseph A. Caprini, M.D.
Chapters 28, 29

John J. Coyle, M.D.
Chapters 9, 35, 47

Daphne W. Denham, M.D.
Chapters 51-53

Woody Denham, M.D.
Chapters 18, 19, 21

Gregory Dumanian, M.D.
Chapter 71

Mark K. Eskandari, M.D.
Chapters 97, 103, 104, 106, 107

Julius W. Few, Jr., M.D.
Chapters 69-70

Neil A. Fine, M.D.
Chapters 73, 74

James W. Frederiksen, M.D.
Chapters 78, 79

Jonathan Fryer, M.D.
Chapter 91

Sean C. Grondin, M.D.
Chapter 1

Amy L. Halverson, M.D.
Chapters 39, 40, 43, 56

Keith A. Horvath, M.D.
Chapter 83

Raymond J. Joehl, M.D.
Chapters 2, 4

Seema A. Khan, M.D.
Chapter 62

Alan J. Koffron, M.D.
Chapters 58, 84

Thomas W. Kornmesser, M.D.
Chapters 102, 105

Joseph R. Leventhal, M.D., Ph.D.
Chapters 85, 86

Michael J. Liptay, M.D.
Chapter 3

Jon S. Matsumura, M.D.
Chapters 94-96

Mark D. Morasch, M.D.
Chapter 93

Monica Morrow, M.D.
Chapters 61, 63

Joseph P. Muldoon, M.D.
Chapters 41, 42

Kenric M. Murayama, M.D.
Chapters 15, 17, 34, 48

Alexander P. Nagle, M.D.
Chapter 30

Jay B. Prystowsky, M.D.
Chapters 8, 14, 16

Nancy Schindler, M.D.
Chapters 108-110

Joseph R. Schneider, M.D., Ph.D.
Chapters 98, 99

Stephen F. Sener, M.D.
Chapter 6

Valeric L. Staradub, M.D.
Chapters 64, 65

Steven J. Stryker, M.D.
Chapters 32, 38, 44, 57

Sudhir Sundaresan, M.D.
Chapters 75-77

Mark Toyama, M.D.
Chapters 26, 31, 46

Robert Vanecko, M.D.
Chapter 81

Jeffrey D. Wayne, M.D.
Chapters 7, 55, 66-68

Michael A. West, M.D., Ph.D.
Chapters 12, 27, 49

John V. White, M.D.
Chapter 100

David J. Winchester, M.D.
Chapters 33, 36

Preface

Performing a surgical operation must be one of the most complex motor tasks undertaken by humans. To the student rotating on a third-year clerkship, or to the resident beginning a surgical career, the complexity can no doubt be daunting. It is the goal of this handbook to try to bring some degree of order to the complexity, focusing on many of the common operations in general, plastic, thoracic, and vascular surgery, and in organ transplantation. We have done so by dividing each procedure into specific and well-defined steps.

Research originating in the 1950s explored whether teaching a novice to perform a skill is best accomplished by having the learner perform the entire skill (called the "whole training method") or first practice the parts of the skill (called the "part-whole method"). It was concluded that certain types of skills were more efficiently learned using the latter method. As novices gain proficiency in the various steps (or subroutines) of a skill, their ability to more fluidly chain the steps together increases. Learning complex skills by practicing its sub-skills continues to be supported today by educational learning theories and practices.

A master surgeon brings flow and continuity to an operation which may make the procedure appear to be an indivisible whole, but in fact all operations are composed of a series of steps. Mastery comes from practicing each step repeatedly until competence is obtained and then integrating the steps into a whole. Accomplished surgeons not only do each step well, but also have a clear mental image of the progression of the steps and their interdependency.

Unlike the music student, who may study with a single teacher for an extended period of time, medical students and residents often scrub with several surgeon-teachers, all of whom have their particular approach to a given operation. For the advanced student, such variety of experience may be enriching, but for the novice, the variety of approaches can be confusing. In this book, we hope to provide the beginner with a framework that can serve as a baseline and against which real-life experiences can be measured.

One of our goals in creating this book was to make it portable, so that it could be used anywhere when a few minutes were available. Landes Bioscience has made this possible and created an attractive and convenient format for the material.

We would be remiss if we did not acknowledge the excellent illustrations created by Simon Kimm. We wanted to capture some essential visual features to accompany the text, and Simon has done this admirably.

We hope that you enjoy this book and that it helps you better envision and understand the complex and beautiful art of surgery.

Richard H. Bell, Jr., M.D.
Dixon B. Kaufman, M.D., Ph.D.
Chicago, Illinois
April 2005

Section 1: Gastrointestinal

Section Editors: Richard H. Bell, Jr.
and David P. Winchester

Chapter 1

Esophageal Diverticulectomy: Zenker's

Sean C. Grondin

Indications

A Zenker's (pharyngoesophageal) diverticulum is an acquired pulsion diverticulum that develops secondary to cricopharyngeal dysfunction. The diverticulum, which typically enlarges with time, arises from a triangular weakening (Killian's triangle) in the posterior midline of the lower pharynx between the transverse and oblique muscle fibers of the cricopharyngeus muscle (inferior constrictor). Operative intervention should be considered in symptomatic patients.

Preop

An upper gastrointestinal X-ray and careful endoscopy are essential to confirm the diagnosis of Zenker's diverticulum. Esophageal manometry is reserved for patients suspected of having associated foregut disorders. Patients with associated gastroesophageal reflux disease (25%) should be treated adequately preoperatively. General anesthesia is performed with single-lumen endotracheal intubation. Patients with large diverticula should be kept on clear fluids 2-3 days prior to surgery to decrease the risk of aspiration during induction of anesthesia. Prophylactic antibiotics and deep vein thrombosis prophylaxis are routinely used.

Procedure

Step 1. Place patient in supine position with arms at the sides and insert a shoulder roll to extend the neck.

Step 2. Place a nasogastric tube into the stomach using an esophagogastroscope to prevent perforation of the diverticulum from a "blind" insertion.

Step 3. Make a longitudinal incision along the anterior border of the left sternocleidomastoid muscle. Incise the platysma muscle and divide the omohyoid muscle. Gently retract the carotid artery, internal jugular vein, and vagus nerve laterally allowing the retroesophageal area to be developed with blunt dissection. If necessary, divide the middle thyroid vein.

Step 4. Palpate the nasogastric tube and encircle the esophagus with a Penrose drain.

Step 5. Locate the diverticulum posteriorly, just inferior to the cricoid cartilage. Grasp the diverticulum gently with a Babcock forcep. Using blunt and sharp dissection, identify the neck of the diverticulum. Remove the nasogastric tube and place a Maloney bougie (48-54 F) down the esophagus.

Step 6. Divide the cricopharyngeus muscle (cricomyotomy) 3-4 cm distally and 2 cm proximally.

Northwestern Handbook of Surgical Procedures, edited by Richard H. Bell, Jr. and Dixon B. Kaufman. ©2005 Landes Bioscience.

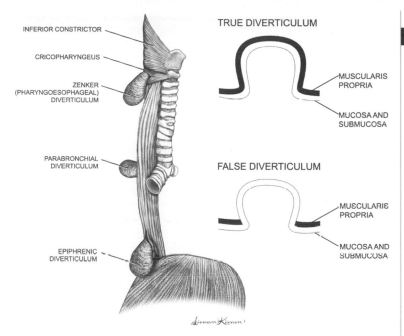

Figure 1.1. Esophageal diverticula.

Step 7. Excise diverticula transversely using a mechanical stapler device or by standard suture technique. Alternatively, suspend the diverticulum from the prevertebral fascia (diverticulopexy). Endoscopic treatment of the diverticulum is controversial.

Step 8. Remove the bougie and gently place an esophagogastroscope into the proximal esophagus. Insufflate the esophagus to identify any unrecognized perforation or a staple line leak that requires repair.

Step 9. Place a Jackson-Pratt drain in the retroesophageal space. Reapproximate the deep cervical fascia with interrupted absorbable suture. Close the subcutaneous tissue and skin with a continuous absorbable suture.

Postop

Keep patient NPO and instruct hospital staff not to place a nasogastric tube without endoscopic assistance. Perform a gastrograffin or dilute barium swallow 1-2 days postoperatively to rule out a leak from the myotomy or diverticulectomy site. If no leak is detected, start the patient on clear fluids and advance diet over 2-3 days.

Complications

Esophageal leak, inadequate myotomy, esophageal stricture, and recurrent laryngeal nerve injury.

Follow-Up

Dietary counseling and treatment of associated gastroesophageal reflux disease.

Chapter 2

Heller Myotomy: Laparoscopic

Raymond J. Joehl

Indications

Laparoscopic Heller myotomy is indicated in operable patients who are candidates for laparoscopy and who have had diagnostic studies indicating a swallowing disorder consistent with achalasia, especially a barium swallow (showing the characteristic bird-beak appearance of the narrowed esophagogastric junction (EGJ)), an upper endoscopy to rule out the possibility of a neoplasm (pseudoachalasia), and esophageal manometry that shows absence of peristalsis in the proximal and midesophagus and absence of swallow-induced relaxation of the lower esophageal sphincter (LES), usually with high basal pressure >20 mmHg.

Preop

Preoperative studies should include barium swallow, esophagogastroduodenoscopy (EGD), and esophageal manometry.

Procedure

Step 1. The patient is placed supine in a modified lithotomy position with the operating table rotated 20^0 clockwise (to provide anesthesia access to patient's head around/beneath monitor). The surgeon stands between the patient's legs and assistant stands at the left side.

Step 2. After prepping and draping, CO_2 pneumoperitoneum is established and five 5-10 mm laparoscopic ports are placed in the upper abdomen: one in the supraumbilical region to the left of midline, and four in the subcostal region (one at each anterior axillary line and one at each lateral rectus).

Step 3. A 5-10 mm liver retractor is placed through the far right port in the anterior axillary line, elevating the left hepatic lobe and exposing the proximal stomach and esophageal hiatus. The retractor is secured to a table-mounted retractor holder.

Step 4. The gastrosplenic omentum is divided from the fundus of the stomach to the angle of His with an ultrasonic scalpel, mobilizing the fundus off the left half of the diaphragmatic crus and incising the retroperitoneum behind and at the left side of gastric cardia.

Step 5. The upper portion of the gastrohepatic omentum and the phrenoesophageal membrane over the EGJ are opened, identifying the anterior vagus nerve.

Step 6. In the retroperitoneum medial to the right half of the diaphragmatic crus, the retroesophageal space behind the posterior vagus nerve is opened. The dissection is then connected to the left subphrenic space. A half-inch Penrose drain is passed behind the esophagus in the retroesophageal space to surround the EGJ.

Northwestern Handbook of Surgical Procedures, edited by Richard H. Bell, Jr. and Dixon B. Kaufman. ©2005 Landes Bioscience.

2

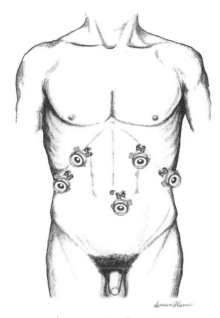

Figure 2.1. Heller myotomy, laparoscopic. Port sites.

The drain is secured with a locking Allis laparoscopic grasping forceps, placing gentle downward traction on the EGJ.

Step 7. The distal esophagus is separated from the apex of crural diaphragm.

Step 8. An upper endoscope is passed to empty the dilated esophagus and identify the squamocolumnar junction at the EGJ by transillumination.

Step 9. The myotomy is begun on the anterior cardia approximately 2 cm distal to the EGJ dissecting into the submucosal plane using a Maryland dissector and either an ultrasonic dissector, electrocautery hook, or 5 mm blunt suction tip.

Step 10. The myotomy is extended retrograde into the submucosal plane of the distal esophagus until the myotomy is 8-10 cm in length.

Step 11. Endoscopy is performed again to assess myotomy completeness by gently blowing air into the esophageal lumen to assess the ease of LES opening; then esophageal manometry is performed across the myotomized EGJ to exclude a persistent high pressure zone.

Step 12. When the myotomy is judged to be complete by endoscopy and manometry, all fluid and air are suctioned from the stomach. A Dor anterior fundoplasty is then performed by folding the fundus of the stomach over the myotomy surface and suturing it to the right edge of the myotomy and to the right half of the diaphragmatic crus using nonabsorbable 2-0 suture and polytetrafluoroethylene (PTFE) pledgets. Three to four interrupted sutures are required to complete the fundoplasty.

Step 13. The upper abdomen and subphrenic space are irrigated with warm saline, inspecting the subphrenic space and spleen for bleeding. The fascia is closed at the sites of 10 mm ports.

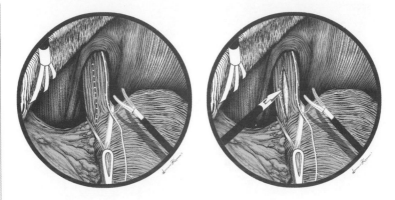

Figure 2.2. Heller myotomy, laparoscopic.

Additional Comments

- If a hiatal hernia is noted, the hiatus may be narrowed with suture(s) placed to approximate the left and right halves of the crus behind the EGJ.
- If esophageal perforation occurs during the myotomy, repair the perforation with interrupted 3-0 silk suture using intracorporeal knot-tying.

Postop

It is not necessary to use a nasogastric tube. Oral intake of liquids and soft foods (with no red meat or dry bread) is begun as soon as the patient has no nausea, usually on the day of surgery. Pain is controlled with injectable narcotics for 6-12 hours and then oral pain medicine. If intraoperative esophageal perforation occurred (5-15% chance), the patient is kept NPO and a water-soluble-contrast X-ray swallow is performed the following morning; if there are no signs of leakage, PO fluids are begun. If leakage is noted, the patient should be monitored closely for clinical signs of esophageal perforation (fever, vital sign changes suggesting sepsis, breathing difficulties, chest pain, subcutaneous emphysema) and a thoracic surgeon consulted. If there are no clinical signs of esophageal perforation, the patient is kept NPO and a repeat water-soluble-contrast X-ray swallow performed in 5-7 days.

Complications

Vagal injury, esophageal perforation, gastric perforation, splenic laceration and hemorrhage, gastroesophageal reflux disease (GERD).

Long-Term Follow-Up

Proton pump inhibitor (PPI) therapy is given for one month, then discontinued; if symptoms of GERD develop, then lifelong daily PPI therapy may be indicated. A timed barium swallow is done in 6 months. At one year, a timed barium swallow, EGD, and manometry are performed. If patients develop persistent dysphagia at any time after Heller myotomy, a full evaluation with barium swallow, esophageal manometry, and EGD is indicated.

Chapter 3

Thoracic Esophageal Perforation Repair

Michael J. Liptay

Indications

1. Boerhaave's esophagus—violent retching against closed glottis causes esophageal blowout usually on left side just above gastroesophageal (GE) junction.
2. Iatrogenic dilatation/biopsy/foreign body extraction.

Preop

Time from diagnosis to operative treatment is critical to the chance for a successful primary repair and avoidance of life-threatening sepsis. Diagnosis by chest X-ray, upper gastrointestinal study (Gastrograffin) or CT scan should be followed by IV hydration, broad-spectrum IV antibiotics, and emergent trip to the operating room.

Procedure

Step 1. Incision in the 8th intercostal space via posterolateral thoracotomy on the side of pleural soilage.

Step 2. Mobilize the inferior pulmonary ligament and open the mediastinal phlegmon partially contained by the pleura.

Step 3. Copious irrigation to wash out detritus in the mediastinum/pleural space.

Step 4. Debride nonviable tissue.

Step 5. Identify mucosal limits of the esophageal tear.

Step 6. If tissue appears healthy (usually the case within 24-36 hours of the insult), attempt primary repair.

Step 7. Have anesthesia pass a nasogastric tube into the stomach beyond the defect under the surgeon's direct vision. Perform a two-layer closure using interrupted 3-0 silk sutures first on the full extent of the mucosal defect. Cover that repair with an interrupted layer in the muscular wall taking both inner circular and outer longitudinal layers together.

Step 8. Consider covering the repair with a parietal pleural flap to provide additional support.

Step 9. Insert two chest tubes anterior and posterior to the repair and close.

Addsteps. If a primary repair of the esophageal injury is not feasible—usually secondary to autodigestion of the tissue and mediastinal sepsis from delayed diagnosis—then either an esophageal exclusion procedure (staple off GE junction and cervical esophagostomy and jejunostomy feeding tube) or an esophagogastectomy with gastric pullup need to be considered.

Northwestern Handbook of Surgical Procedures, edited by Richard H. Bell, Jr. and Dixon B. Kaufman. ©2005 Landes Bioscience.

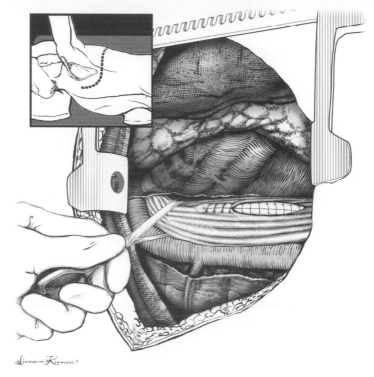

Figure 3.1. Repair of thoracic esophageal perforation.

Postop

Broad-spectrum antibiotics, fluid resuscitation, and total parenteral nutrition are often required for 7-10 days. An upper gastrointestinal study should be planned for 7 days postoperatively.

Complications

The major complication is failed repair with recurrent esophageal leak and mediastinal sepsis.

Follow-Up

Long-term follow-up is geared towards treating the underlying condition (stricture, alcoholism, etc.) and watching for fibrotic stricture development at repair site.

Chapter 4

Antireflux Procedure: Laparoscopic (Nissen)

Raymond J. Joehl

Indications

Nissen fundoplication is indicated in operable patients with objectively documented gastroesophageal reflux disease (GERD) whose symptoms/signs have not been controlled by medical management with proton pump inhibitor (PPI) therapy and other measures.

Preop

Prior to surgery, patients should be fully evaluated by esophagogastroduodenoscopy (EGD) and distal esophageal biopsy, esophageal manometry, 24-hour pH study (if no objective evidence of reflux or if symptoms are atypical), and barium swallow if hiatus hernia is large (>3-4 cm). Immediately prior to surgery, patients should have appropriate deep vein thrombosis (DVT) prophylaxis based on risk factors.

Procedure

Step 1. The patient is positioned supine in modified lithotomy with the bed rotated 20° clockwise. The surgeon stands between the patient's legs and the assistant stands at the left side.

Step 2. After prepping and draping, CO_2 pneumoperitoneum is established and five 5-10 mm laparoscopic ports are placed in the upper abdomen: one in the supraumbilical region to the left of midline and four in the subcostal region (one at each anterior axillary line and one at each lateral rectus).

Step 3. A 5-10 mm liver retractor is placed through the far right port in the anterior axillary line, elevating the left hepatic lobe, thereby exposing the proximal stomach and esophageal hiatus. The retractor is secured to a table-mounted retractor holder.

Step 4. An orogastric (OG) tube is positioned in the midstomach for decompression.

Step 5. The gastrocolic and gastrosplenic omenta are divided along the greater curvature from midstomach up to the angle of His with an ultrasonic scalpel, mobilizing the fundus off the left half of the diaphragmatic crus and opening the retroperitoneum behind and at the left side of the gastric cardia.

Step 6. The upper portion of the gastrohepatic omentum is divided and the phrenoesophageal membrane opened over the esophagogastric junction (EGJ), identifying the anterior vagus nerve.

Step 7. The retroperitoneum medial to the right half of the diaphragmatic crus is entered and the retroesophageal space behind the posterior vagus nerve opened, and this dissection connected to the left subphrenic space, going behind the esophagus.

Northwestern Handbook of Surgical Procedures, edited by Richard H. Bell, Jr. and Dixon B. Kaufman. ©2005 Landes Bioscience.

Step 8. A half-inch Penrose drain is passed through the retroesophageal space to surround the EGJ. It is secured with a locking Allis laparoscopic grasping forceps, placing gentle downward traction on the EGJ.

Step 9. The distal esophagus is mobilized out of the posterior mediastinum and hiatal canal and any hiatus hernia reduced. At this point, the anesthesiologist should remove the OG tube and gently pass a 50-52 Fr esophageal dilator down the esophagus through the EGJ into the stomach.

Step 10. An enlarged hiatal orifice behind the esophagus is eliminated by reapproximating the left and right halves of the crus with interrupted 0 or 2-0 braided polyester sutures over PTFE pledgets.

Step 11. The gastric fundus is passed from left to right behind the EGJ and the stomach circumferentially plicated around the EGJ with at least two anteriorly placed interrupted 2-0 braided polyester sutures, grasping the esophagus to the right of the anterior vagus nerve between bites of fundus. The length of the wrap should be approximately 2 cm.

Step 12. The esophageal dilator is removed. The OG tube is replaced in the mid body of stomach only if the stomach appears distended with air or fluid.

Step 13. The abdomen is irrigated with warm saline, inspecting the subphrenic space and spleen for bleeding, and then aspirating all saline irrigation.

Step 14. The liver retractor, the table-mounted retractor holder, and all laparoscopic ports are removed. The fascia at sites of 10 mm ports is closed. Skin incisions are closed with subcuticular sutures.

Additional Comments

If inadequate esophageal length is obtained after reducing the hiatus hernia in Step 9 and the EGJ cannot be returned to an intraabdominal position, a Collis gastroplasty may be indicated.

Postop

A Foley catheter is necessary only in patients having a long procedure or in whom there is indication to monitor urine output closely. An NG or OG tube is usually not necessary. Oral intake of liquids and soft foods is begun as soon as the patient has no nausea, usually on the day of surgery. Red meat and dry bread are avoided for 2-4 weeks. Pain is controlled with injectable narcotics for 6-12 hours and then with oral analgesics.

Complications

Complications of Nissen fundoplication include vagal injury, esophageal perforation, gastric perforation, splenic laceration and hemorrhage, gastroparesis, gastric bloating, esophageal dysmotility (dysphagia), excess flatulence, and fundoplication dehiscence with recurrent GERD.

Long-Term Follow-Up

A barium swallow and/or esophagogastroduodenoscopy (EGD) should be performed in patients with persistent dysphagia after 4-8 weeks. A postsurgical EGD should be done in 3-6 months in patients operated for erosive esophagitis. In patients with Barrett's esophagus, a follow-up EGD should be done in 6-12 months and then every 6 months to 3 years depending on the presence or absence of dysphagia. Recurrent GERD symptoms require a complete evaluation including a barium swallow, esophageal manometry, EGD, and possibly a 24-hour pH study.

Chapter 5

Gastrostomy: Open

Richard S. Berk

Indications

The indication for open gastrostomy is primarily the inability of the patient to be nourished by the oral route due to obstruction or dysfunction of the oral cavity, pharynx, or esophagus. With the advent of percutaneous endoscopic gastrostomy (PEG), open gastrostomy as an isolated procedure is typically limited to those in whom PEG placement is not possible or is contraindicated. Open gastrostomy tube placement may also be performed as an adjunct to other abdominal procedures in patients with poor respiratory reserve who may not tolerate a nasogastric tube or in patients who are expected to have a long period before they are able to resume oral feedings.

Preop

Patients in need of open gastrostomy should have fluid and electrolyte balance optimized to the extent possible. Coagulation factors may need to be brought to acceptable levels since these patients are often nutritionally depleted by their underlying disease.

Procedure

Step 1. The patient is placed in the supine postion and prepped from above the xiphoid to the pubis, laterally to each side of the abdominal wall. Drapes are placed at the xiphoid, umbilicus, right rectus edge, and left edge of the anterior abdominal wall.

Step 2. An incision is made in the midline below the xiphoid, large enough to admit approximately four fingerbreadths and to visualize the anterior gastric wall.

Step 3. The skin site for the gastrostomy is chosen that is away from the costal margin and overlies the stomach so that the gastric wall can be easily brought up and attached to the abdominal wall.

Step 4. A 0.5-1.0 cm incision is made in the skin and a 6" clamp placed through the full thickness into the peritoneal cavity, grasping an 8" clamp and bringing the latter back outside the peritoneal cavity.

Step 5. The tip of a 24 F Foley catheter with a 5 cc balloon is then grasped by the 8" clamp and brought back inside the abdominal cavity.

Step 6. Two Babcock clamps grasp the gastric wall on each side of the gastrostomy site and electrocautery is used to make a small opening in the muscular layers. The submucosa and mucosa are grasped with mosquito clamps and the lumen entered, tying the tissue with 4-0 absorbable suture.

Northwestern Handbook of Surgical Procedures, edited by Richard H. Bell, Jr. and Dixon B. Kaufman. ©2005 Landes Bioscience.

5

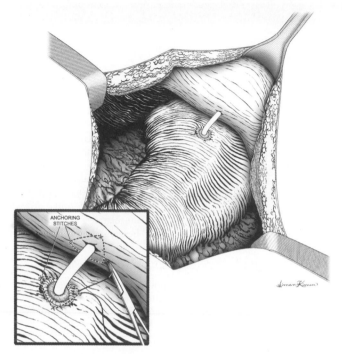

Figure 5.1. Open gastrostomy.

Step 7. The opening is enlarged and the Foley tip is inserted. Two concentric pursestring sutures of 2-0 or 3-0 silk are placed sequentially, tying the inner one to invert the gastric edge, then tying the outer one afterward.

Step 8. Three or four 2-0 silk sutures are then placed adjacent to the openings in the peritoneum and stomach, starting at the point most distant from the operator, and then closer.

Step 9. The Foley balloon is inflated and the stomach is pulled up to the abdominal wall by traction on the tube. The sutures are then tied in the order in which they were placed.

Step 10. The tube is fixed to the skin of the abdominal wall with a 2-0 silk or nylon suture.

Step 11. The abdominal incision is closed with 0 nylon on the fascia, 3-0 absorbable suture on Scarpa's layer, and 4-0 subcuticular sutures, staples, or steri-strips. The gastrostomy tube is attached to a drainage bag, and a tape "mesentery" is attached to the tube to protect it from being pulled out.

Postop

The gastrostomy tube is left open to drainage for 24 hours and may then be used for feeding, if that was its intent, or to continued drainage. Topical povidone may be used daily to prevent skin inflammation from the holding suture. Maintaining a tape mesentery on the tube itself protects against inadvertent pulling on the suture holding the tube in place. If the tube is used for feeding, it may be coiled up on the abdominal wall and taped in place when not in use.

Complications

Complications associated with this procedure include bleeding from the gastric mucosa, dislodgement of the gastrostomy tube, leakage of gastric contents with excoriation of the protective skin surface, and infection at the tube site. Gastric obstruction may occur if the tube is placed too proximal on the gastric wall causing tenting of the stomach or if the tube migrates distally and the balloon occludes the pylorus.

Follow-Up

Long-term follow-up for open gastrostomies involves maintaining of the tube lumen and preventing infection at the tube site. Changing the tube is easily accomplished by deflating the balloon, removing the tube after cutting the suture, and replacing it immediately. If the tube comes out inadvertently at home, immediate return for replacement is mandatory; in certain cases, the patient can be taught to replace the tube.

Chapter 6

Gastrectomy: Total

Stephen F. Sener and Malcolm M. Bilimoria

Indications

Total gastrectomy is indicated in patients with biopsy-proven cancer of the proximal stomach or antral cancer that extends to the proximal stomach, who are candidates for general anesthesia. Regional lymph node involvement which can be encompassed in the resection is acceptable but distant metastases are a contraindication to total gastrectomy. Some patients with nongastric cancers that involve the stomach (i.e., retroperitoneal sarcomas) may also be candidates for total gastrectomy.

Preop

Patients should be preoperatively staged with a chest X-ray, CT scan of the abdomen and pelvis, and esophagogastroduodenoscopy. Standard prophylactic antibiotics are administered 30 minutes preoperatively. A thoracic epidural cathether is placed for pain management. A nasogastric tube and a urinary catheter should be inserted prior to making the incision. Sequential compression boots and subcutaneous heparin are generally indicated for prophylaxis and should be started preoperatively.

Procedure

Step 1. The patient is placed in the supine position. Either a midline or Chevron incision can be used. A small incision can be used to perform a preliminary exploration of the abdomen and then enlarged as necessary if there are no findings which would obviate resection. A self-retaining retractor should be inserted to provide good exposure of the esophageal hiatus and celiac lymph nodes.

Step 2. All peritoneal surfaces are inspected and intraoperative ultrasound of the liver performed to rule out metastatic disease.

Step 3. The omentum is completely mobilized off the transverse colon so that it can be removed en bloc with the stomach. The posterior stomach is examined through the lesser sac. If the primary tumor invades the short gastric vessels or the splenic capsule, splenectomy will be necessary.

Step 4. A Kocher maneuver of the duodenum is performed. The subpyloric nodes are separated from the pancreatic head, and the pyloric vein is ligated at its junction with the middle colic vein. The right gastric artery and veins are ligated just distal to the pylorus. The duodenum just distal to the pylorus is transected between a linear stapler and a short Kocher clamp. The stapled duodenal stump is oversewn with 3-0 silk.

Step 5. With the greater omentum retracted in a cephalad direction, the dissection is continued along the greater curvature of the stomach up into the short gastric vessels. If a splenectomy is not necessary, the short gastric vessels are ligated up past the spleen to the gastroesophageal junction. If a splenectomy is required, the splenic

Northwestern Handbook of Surgical Procedures, edited by Richard H. Bell, Jr. and Dixon B. Kaufman. ©2005 Landes Bioscience.

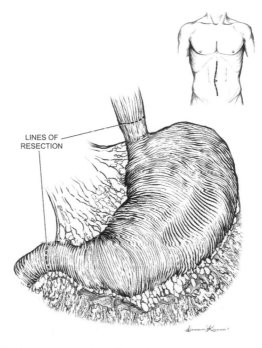

Fligure 6.1. Total gastrectomy. Boundary of resection.

artery is ligated near its origin from the celiac axis and the splenic artery lymph nodes are taken in continuity with the spleen and stomach. The splenic vein is ligated and divided in the splenic hilum.

Step 6. The stomach is then retracted cephalad to expose the left gastric artery, the celiac lymph nodes, and the aorta. The left gastric artery is ligated at its junction with the celiac axis. The common hepatic artery lymph nodes are dissected off the artery and retracted medially into the left gastric artery lymph node bundle. If a splenectomy was not done, the splenic artery lymph nodes are dissected away from the artery and retracted in the left gastric artery lymph node bundle. The paraaortic lymph nodes are separated from the aorta and retracted up into the left gastric artery lymph node bundle, which is attached to the gastrectomy specimen. This maneuver completes the R-2 lymph node dissection.

Step 7. An automatic pursestring applier is then placed across the esophagus approximately 2 cm above the proximal margin of gross tumor or the gastroesophageal junction, whichever is more proximal. The esophagus is transected, and the gastrectomy specimen is sent to pathology for immediate analysis of the esophageal margin.

Step 8. Approximately 10 cm distal to the ligament of Treitz, the jejunum is mobilized and transected at that location using a linear cutter. The mesentery is divided vertically as necessary to gain mobility. The distal portion of the jejunum is brought through the bare area of the transverse mesocolon to the left of the midline.

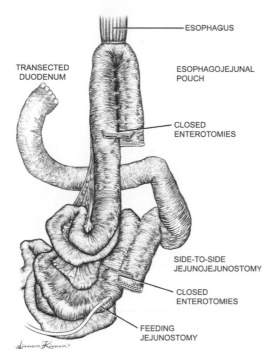

ESOPHAGUS

TRANSECTED
DUODENUM

ESOPHAGOJEJUNAL
POUCH

CLOSED
ENTEROTOMIES

SIDE-TO-SIDE
JEJUNOJEJUNOSTOMY

CLOSED
ENTEROTOMIES

FEEDING
JEJUNOSTOMY

Figure 6.2. Total gastrectomy. Completed reconstruction.

Step 9. The diameter of the distal esophagus is estimated, using the sizers which are provided with circular stapling instruments. An end-to-side esophagojejunal pouch anastomosis should be at least 25 mm in diameter to reduce the risk of postoperative stricture. The circular stapler is separated, the anvil is positioned in the esophagus, and the pursestring suture is tied down around the post of the anvil. The circular stapler is then placed into the end of the jejunal loop by removing the staple line. The trocar of the stapler is advanced through the jejunal wall at a convenient location and is attached to the post of the anvil. The handle of the circular stapling device is rotated to bring the stapling rings together. The tissue is then inspected from the outside to make sure that there is complete approximation of the esophagus and jejunal pouch, and the device is fired. The handle of the instrument is then rotated so that the anvil is separated from the base by about 1 cm. The device is then rotated in-position, and the posterior lip of the anvil is retracted out through the suture line. After removing the stapling instrument from the pouch, the circular staple line is inspected from the outside of the pouch to establish that the tissues are approximated. Defects in the suture line may be reinforced with interrupted 3-0 silk sutures. The tissue rings within the stapler are inspected to ensure that they are intact. The open end of the jejunum through which the stapler has been inserted is reclosed using a linear stapler or a two-layer hand-sewn closure. The nasogastric tube is repositioned so that it passes through the esophagojejunal anastomosis.

Alternative Method for Esophagojejunostomy: Hunt-Lawrence Pouch

Approximately 10 cm of the jejunum is doubled back onto itself to create an inverted "J" configuration. Two small punctures are made side-by-side in the two adjacent limbs of the "J," and a linear cutter is used to create the pouch. Once this is done, the surgeon puts the index finger into the pouch as far as possible through the common enterotomy and loops the finger around the remaining web of common jejunal wall that still exists in the proximal pouch. A second firing of the linear cutter transects the residual web. The circular stapler is then placed into the pouch through the distal enterotomy and the esophagojejunostomy performed as above. The enterotomy through which the circular stapler is introduced is closed with a linear stapler.

Step 10. 50 to 60 cm below the esophagojejunostomy, a side-to-side jejunojejunostomy or functional stapled end-to-side jejunojejunostomy is constructed to re-establish intestinal continuity.

Step 11. A feeding jejunostomy may be placed for postoperative nutrition.

Step 12. The abdomen is closed in layers.

Postop

Patients spend the first postoperative night in the intensive care unit with emphasis on ambulation and pulmonary toilet each day after that. A Gastrograffin upper gastrointestinal study is done about 5 days after the operation. Once it is established that the esophagojejunal anastomosis does not leak, the nasogastric tube may be removed.

Complications

A leak at the esophagojejunostomy site may manifest clinically with tachycardia, fever, pain, and elevated white blood cell count. The most common time for a leak to occur is approximately one week postoperatively.

Follow-Up

Because cancer recurrence is quite frequent after gastrectomy, patients should be followed indefinitely postoperatively with careful attention to symptoms that might suggest recurrence. There is no advantage to routine imaging or laboratory tests in patients who are asymptomatic.

Chapter 7

Gastrectomy: Subtotal or Partial

Jeffrey D. Wayne and Richard H. Bell, Jr.

Indications

Subtotal gastrectomy is indicated for the treatment of gastric adenocarcinoma in the absence of distant metastases. In a subtotal gastrectomy for cancer, it is typical to remove about 85% of the stomach, leaving only a small fundic remnant attached proximally to the esophagus. Subtotal or partial gastrectomy may be indicated for other less common gastric neoplasms, such as stromal tumors.

Partial gastrectomy is also indicated for the treatment of gastric ulcer. In this case, the goal is removal of the gastric antrum only, and approximately 40% of the stomach is typically removed. No vagotomy is performed.

Finally, partial gastrectomy (antrectomy) with truncal vagotomy is occasionally indicated in the treatment of severe or recurrent duodenal ulcer, although the operation is performed much less commonly than a few decades ago.

In all cases of partial or subtotal gastrectomy, the distal margin of resection should be in the duodenum, approximately 1-2 cm beyond the pylorus.

Preop

An intravenous dose of second-generation cephalosporin should be given about 30 minutes prior to incision. Antiembolism prophylaxis should be employed, using subcutaneous heparin and/or sequential compression boots depending on individual risk factors.

Procedure

Step 1. Under general anesthesia, a midline incision is made from the xiphoid to below the umbilicus. A bilateral subcostal incision may also be used if the patient has a shallow costal angle. The subsequent course of the operation varies depending on the indication (cancer or benign ulcer), so the steps are described separately below.

Subtotal Gastrectomy for Cancer

Step 2. If operating for cancer, the omentum should be dissected away from the transverse colon along its entire length and the entire omentum mobilized superiorly en bloc with the stomach.

Step 3. A Kocher maneuver is performed to mobilize the duodenum.

Step 4. The right gastroepiploic vessels are divided beyond the pylorus. The vessels should be ligated as far away from the gastric wall as possible so that any accompanying nodes can be swept toward the stomach and included in the specimen.

Step 5. The right gastric vessels should be divided at the left edge of the porta hepatis, as the right gastric artery arises from the proper hepatic artery. Again, any visible nodes should be kept with the specimen.

Northwestern Handbook of Surgical Procedures, edited by Richard H. Bell, Jr. and Dixon B. Kaufman. ©2005 Landes Bioscience.

Step 6. The duodenum is circumferentially dissected about 2 cm distal to the pylorus and divided with a linear stapler. The staple line on the distal duodenal stump is imbricated beneath 3-0 Lembert sutures.

Step 7. Holding the stomach up and to the patient's left, the lesser omentum is divided close to the liver, ultimately exposing the left gastric artery as it sweeps up onto the lesser curvature of the stomach.

Step 8. The left gastric artery is divided and ligated at its base as it arises from the celiac axis. A suture ligature should be used on the stump of the artery. The accompanying vein(s) is also divided and ligated. Again, all soft tissue and nodes running along the left gastric vessels are included in the specimen.

Step 9. Any remaining short gastric vessels on the greater curvature of the stomach are ligated and divided up to the level of the gastroesophageal junction.

Step 10. The stomach is divided using two firings of a linear stapler. The first stapler is applied at approximate right angles to the greater curvature, and the length of the first cut is designed to be the length of the planned gastrojejunal anastomosis. The second firing of the stapler should angle up to the top of the lesser curvature of the stomach, within 1-2 cm of the esophagus. With the second firing of the stapler, the specimen is removed. The second staple line is then oversewn with running or interrupted suture.

Step 11. The proximal jejunum is brought up to the stomach in either an antecolic or retrocolic position, depending on how the organs best lie. A back row of interrupted sutures is placed between the jejunum and the posterior stomach just behind the existing first staple line. Once the posterior wall is complete, the first gastric staple line is excised and a matching or slightly smaller opening is made in the jejunum. The inner layer of the anastomosis is then performed using a running suture, beginning on the back wall and coming around the corners to the front. The anastomosis is completed with an anterior row of interrupted Lembert sutures.

Step 12. If the jejunum was passed in a retrocolic position, the defect in the transverse mesocolon should be closed to prevent internal hernias. The abdominal wall is then closed with nonabsorbable sutures on the fascia and staples on the skin.

Operation for Ulcer Disease

If an antrectomy is being performed for gastric ulcer or a vagotomy and antrectomy is being performed for duodenal ulcer, the dissection is usually begun proximally and extends distally towards the duodenum, the opposite approach from a cancer operation. The reason for this difference is that the duodenum may be quite scarred in an ulcer operation, and dividing the stomach proximally allows the distal stomach to be lifted up and manipulated, providing better circumferential exposure of the duodenum.

Step 2. If a truncal vagotomy is to be performed, the left triangular ligament of the liver is divided and the left lateral liver segment retracted to the right to expose the gastroesophageal junction. The peritoneum overlying the gastroesophageal junction is incised.

Step 3. The distal esophagus is encircled with a finger, working from left to right behind the esophagus. By pulling down on the esophagus, the left (anterior) vagus nerve is stretched and can be felt easily on the anterior surface of the esophagus. It is picked up with a nerve hook and a 2 cm length of the nerve cleaned. Two medium clips are placed on the nerve and a 1 cm section of nerve between the clips excised.

Step 4. A Penrose drain is placed around the esophagus and used to retract the gastroesophageal junction inferiorly and slightly to the patient's left. The right index finger is passed behind the esophagus and the right (posterior) vagus nerve identified by palpation as a thick band in the tissue between the aorta and the esophagus. The finger is used to push the right nerve up into view, where it can be grasped with a nerve hook, and a 1 cm section of nerve is excised between staples as described for the left vagus. The sections of both left and right nerve should be sent for frozen-section examination to confirm nerve tissue in both specimens.

Step 5. The proximal extent of the antrum on the lesser curvature of the stomach is estimated by looking for the point where the anterior nerve of Latarjet fans out from the lesser omentum over the anterior surface of the stomach (the so-called crow's foot). The lesser curve should be cleaned 1-2 cm proximal to this point. A point on the greater curvature is also cleared that is approximately halfway up the stomach, usually at the lower end of the short gastric vessels. The stomach is then divided with two fires of a linear cutting stapler, just as described above. The second staple line is oversewn.

Step 6. The distal stomach is grasped and used as a handle to facilitate further dissection. Both the greater curve and the lesser curve of the stomach are then skeletonized until the dissection reaches a point about 2 cm beyond the pylorus. In contradistinction to the cancer operation, all of this dissection may be right along the gastric wall. There is no need to include omentum or nodal tissue. The duodenum is then divided with another firing of the linear cutting stapler and the specimen removed.

Step 7. Reconstruction of the gastrointestinal tract can either be done by sewing the gastric remnant to the duodenum (Billroth I) or to the jejunum (Billroth II). Because less stomach is removed in ulcer operations than in cancer operations, a Billroth I reconstruction is often possible, whereas this is almost never the case after an adequate cancer operation. When operating for ulcer disease, a Billroth I reconstruction is probably preferable if it can be done without tension. The anastomosis can be done by placing a posterior row between the stomach and duodenum, then removing the staple line from the duodenum and the first staple line from the stomach, and placing an inner running suture. The anastomosis is completed with an anterior row of interrupted Lembert sutures. If a Billroth II reconstruction is chosen, it is performed as described for the cancer operation above.

Postop

A nasogastric tube is typically placed after surgery and removed when there are signs of bowel activity.

Complications

Splenic injury may occur because of inadvertent traction during the dissection of the stomach or the performance of vagotomy. The injury may not be recognized during surgery. Signs of significant blood loss should prompt reexploration. The duodenal "stump" may leak after a Billroth II reconstruction and require reoperation. Internal hernias may occur through the mesocolon after retrocolic reconstruction. Occasionally, patients may develop bile reflux gastritis or afferent loop syndrome. Recurrent ulcer is very rare. Finally, an aberrant left hepatic artery arising from the left gastric artery may be divided if it is not recognized when dividing the lesser omentum. This may cause ischemia or necrosis of the left hepatic lobe.

Follow-Up

Patients should be seen two or three times in the early postoperative period.

After an operation for cancer, patients should be followed regularly because the incidence of recurrent cancer is high. In addition, referral to a medical oncologist is made, as combined, 5-Fu-based chemoradiation has now been shown to provide a survival advantage when compared to surgery alone in all but the earliest stage of gastric cancer. A change in symptoms may be a harbinger of recurrence and should prompt radiologic imaging and/or endoscopy. Initial follow-up is at 4-month intervals.

Patients operated for ulcer disease do not require antiulcer medications. They do not need endoscopic follow-up unless they develop recurrent symptoms.

Chapter 8

Gastric Bypass: Roux-en-Y

Jay B. Prystyowsky

Indications

Gastric bypass is an operation for morbid obesity and is indicated in patients with a body mass index >40 kg/m^2 or >35 kg/m^2 with significant obesity-related illnesses. The patient should be an appropriate surgical risk and understand the lifestyle changes that the operation requires.

Preop

The patient should be evaluated by a multidisciplinary team and understand the dietary changes that will occur postoperatively. On the day of surgery, antibiotic prophylaxis should be provided and antithrombotic measures instituted, including sequential compression devices and subcutaneous heparin.

Procedure

Step 1. With the patient supine, the abdomen is entered through an upper midline incision; a self-retaining retractor system is placed. The patient is then placed in reverse Trendelenburg position.

Step 2. The left triangular ligament of the liver is divided to allow retraction of the left lateral segment of the liver to the patient's right. At this point the gastroesophageal (GE) junction is mobilized and the distal esophagus elevated anteriorly with a Penrose drain.

Step 3. The stomach is transected with a linear stapler, beginning about 6 cm distal to the GE junction along the lesser curvature and concluding about 2 cm from the GE junction along the greater curvature of the stomach.

Step 4. The small bowel is divided about 40-50 cm below the ligament of Treitz with a linear stapler. A 75-150 cm Roux-en-Y limb is created by performing a handsewn end-to side or a stapled functional end-to-side small bowel anastomosis and closure of the mesenteric defect.

Step 5. The Roux limb is brought into the left upper quadrant in either an antecolic or retrocolic position.

Step 6. An end-to-side gastrojejunostomy with a 21 mm EEA stapler or a handsewn gastrojejunostomy is performed.

Step 7. The small bowel is temporarily occluded beyond the gastric anastomosis to test for anastomotic leaks with instillation of dye or air into the stomach; endoscopy may also be used.

Step 8. Drain placement around the anastomosis is optional.

Step 9. Any defect in the transverse mesocolon is closed if the retrocolic approach was used for the jejunal limb.

Step 10. The midline fascia and skin are closed.

Northwestern Handbook of Surgical Procedures, edited by Richard H. Bell, Jr. and Dixon B. Kaufman. ©2005 Landes Bioscience.

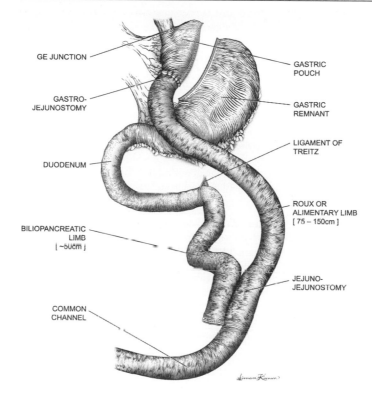

Figure 8.1. Roux-en-Y gastric bypass.

Postop

The use of a nasogastric tube is optional. On postoperative day 1, the patient is given 30 cc water per hour. On postoperative day 2, a Gastrograffin swallow may be performed to rule out a leak; if the study is negative, then the patient is advanced to a sugar-free clear liquid diet.

Patients should ambulate early and can ordinarily be discharged home on postop day 2-4.

Complications

Complications associated with the procedure include anastomotic leak, intraabdominal abscess, splenic injury, anastomotic stricture, wound infection, and wound dehiscence.

Follow-Up

The patient should have frequent follow-up in the first 3-6 months to aid in adjustment to the new diet, then less frequently to assess nutritional status; expected weight loss is about two thirds of excess weight (defined as preoperative weight—ideal body weight).

Chapter 9

Truncal Vagotomy and Pyloroplasty

John J. Coyle

Indications

Truncal vagotomy and pyloroplasty is indicated for the treatment of complicated duodenal ulcer disease. It is the standard operation for bleeding duodenal ulcer. It is less often used for perforation or obstruction. It is also an operative option in patients with ulcers refractory to medical treatment.

Preop

The operation should rarely if ever be done without documentation of a duodenal ulcer by endoscopy or barium studies. In cases of bleeding, large-bore IVs should be placed and aggressive volume resuscitation begun. A nasogastric tube should be placed. A Foley catheter should be placed for monitoring volume status in emergency cases. A prophylactic antibiotic should be given IV 30 minutes prior to incision. Sequential compression devices should be used for deep venous thrombosis prophylaxis.

Procedure

Step 1. A midline abdominal incision is made from the xiphoid process to the umbilicus or slightly below the umbilicus. The incision should be carried up along the side of the xiphoid process to allow maximal exposure of the gastroesophageal (GE) junction area.

Step 2. After entering the abdomen, the left triangular ligament of the liver is incised and the left lateral segment retracted to the patient's right to reveal the GE junction. The peritoneum over the GE junction is opened just below the esophageal hiatus of the diaphragm.

Step 3. The distal esophagus is mobilized circumferentially by blunt dissection and the esophagus surrounded with a Penrose drain.

Step 4. Using the Penrose drain to provide downward traction on the stomach, the anterior and posterior vagal trunks are identified. Caudad retraction of the stomach is often helpful in identifying the nerve trunks by palpation. The anterior nerve crosses the anterior surface of the GE junction. The posterior nerve lies to the right posterior of the esophagus and is a short distance away from the esophagus..

Step 5. Each nerve trunk is isolated in turn with a nerve hook and lifted into view.

Step 6. Metal clips are applied twice to each nerve and a portion of each nerve 5-10 mm long is resected between the clips. Pathologic evaluation of the specimens by frozen section should be done to confirm that they contain nervous tissue.

Step 7. The duodenum is then mobilized liberally with a Kocher maneuver.

Northwestern Handbook of Surgical Procedures, edited by Richard H. Bell, Jr. and Dixon B. Kaufman. ©2005 Landes Bioscience.

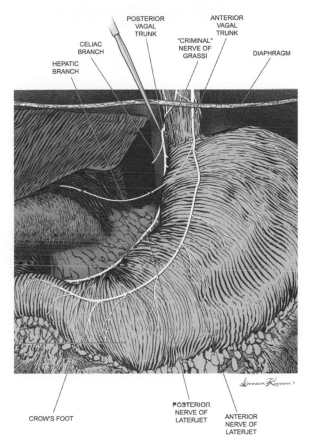

Figure 9.1. Truncal vagotomy.

Step 8. Incise the anterior stomach and duodenum at least 2 cm proximally and distally from the pylorus along the long axis of the bowel. Include perforation in the incision when present.

Step 9. Close resultant defect along transverse axis of the bowel.

Postop
Nasogastric suction until evidence of bowel function.

Complications
Subhepatic abscess, wound infection, gastric atony, dehiscence of pyloroplasty, esophageal perforation.

Follow-Up
Assessment of gastric function, ulcer recurrence.

Chapter 10

Highly Selective (Parietal Cell) Vagotomy

Richard H. Bell, Jr.

Indications

Parietal cell vagotomy is indicated for the treatment of duodenal ulcer disease which does not respond to management with antibiotics and/or acid-reducing medications. The operation should not be used for pyloric channel ulcers due to a high incidence of recurrence. The operation is also indicated as an adjunct to omental closure of acutely perforated duodenal ulcer.

Preop

An intravenous antibiotic for surgical site infection prophylaxis (cefazolin 1 g IVPB) is given 30 minutes before incision. A nasogastric (NG) tube and urinary catheter are placed after general anesthesia is induced. Pneumatic calf compression boots and/or low-dose subcutaneous heparin are used for deep vein thrombosis prophylaxis depending on the patient's risk profile.

Procedure

Step 1. An upper midline abdominal incision is used and the abdomen explored to the extent possible. The position of the NG tube is confirmed.

Step 2. An upper abdominal retractor system is placed. The triangular ligament of the left lateral segment of the liver is divided with cautery, taking care to avoid damaging the inferior phrenic vein. The liver is then retracted to the patient's right to expose the gastroesophageal (GE) junction.

Step 3. The peritoneum over the GE junction is incised, freeing the left lateral margin of the esophagus. An index finger is passed behind the esophagus from the splenic side to the hepatic side. The posterior vagus nerve is identified by palpation as it courses in the soft tissues near the right posterolateral surface of the esophagus. The nerve is elevated with a nerve hook and a vessel loop passed around the nerve.

Step 4. The anterior vagus nerve is identified as it courses on the anterior surface of the esophagus. A vessel loop is passed around the nerve.

Step 5. Moving to the gastric antrum, the point is identified at which the anterior nerve of Latarjet (the gastric branch of the vagus) begins to splay out like a bird's foot over the gastric antrum, usually about 7 cm above the pylorus. Just proximal to this point, the anterior layer of the gastrohepatic omentum is opened between the nerve of Latarjet and the edge of the lesser curvature of the stomach (see Fig. 9.1).

Step 6. Moving proximally towards the GE junction, all of the tissue in the anterior leaflet of the gastrohepatic omentum between the nerve of Latarjet and the gastric wall is divided. Small tissue bites are taken to avoid hematomas, which may make the dissection very difficult. Nearing the GE junction, the position of

Northwestern Handbook of Surgical Procedures, edited by Richard H. Bell, Jr. and Dixon B. Kaufman. ©2005 Landes Bioscience.

encircled anterior vagus nerve is checked to avoid injuring the trunk of the nerve. The dissection is continued until the vessel loop is reached.

Step 7. Returning to the antral area, the same dissection is repeated on the posterior leaflet of the gastrohepatic omentum until reaching the vessel loop around the posterior vagus nerve.

Step 8. A Penrose drain is passed around the distal esophagus with the posterior vagus nerve outside the drain. Retracting the posterior vagal nerve with the vessel loop and using the Penrose drain to manipulate the esophagus, the esophagus is completely cleaned circumferentially for a distance of 5-7 cm. A separate branch of the vagus on the splenic side of the esophagus (the nerve of Grassi) may be present. If it is encountered, it is divided, removing a 1 cm section of the nerve. The Penrose drain is then removed.

Step 9. The spleen is examined to be sure that there has been no inadvertent injury. The lesser curvature of the stomach is reexamined for any bleeding.

Step 10. The midline fascia and skin are closed with suture material of choice.

Postop

The NG tube is removed on postoperative day 1. Diet is instituted as tolerated. Antiulcer medications do not need to be restarted. In cases of emergency operation, the patient should be evaluated for *H. pylori* infection postoperatively if testing has not previously been done.

Complications

Intraoperative injury to the spleen is a potential complication, but is quite unusual. There is little procedure-specific morbidity.

Follow-Up

Patients need not continue ulcer medications. They should be followed indefinitely at regular intervals (6-12 months) to assess for ulcer recurrence. Pain, anemia, or other findings should prompt repeat endoscopy.

10

Chapter 11

Perforated Duodenal Ulcer Repair: Omental Patch

Ermilo Barrera, Jr.

Indications

Omental patch repair is indicated for perforated duodenal ulcers.

Preop

Patients with perforated ulcers may be dehydrated, so intravenous fluid resuscitation should be undertaken. Prophylactic intravenous antibiotics are indicated. Patients are generally begun on H2 receptor antagonists or proton pump inhibitors.

Procedure

Step 1. The patient is placed in the supine position; an upper midline incision is made.

Step 2. Any intraabdominal fluid should be collected in a syringe for culture and sensitivity, then any remaining fluid aspirated..

Step 3. Place 2-3 seromuscular sutures of 2-0 or 3-0 silk so that the sutures cross the perforation but are placed in relatively healthy tissue. Do not tie the sutures yet.

Step 4. Cover the perforation with a tongue of omental fat and tie the previously placed sutures over the omentum.

Step 5. Irrigate copiously and close fascia. Consider the use of retention sutures for elderly patients, poorly nourished patients, immunocompromised patients, etc.

Postop

Remove nasogastric tube when bowel activity resumes. Advance diet as tolerated. Switch from IV to oral antiulcer medications before discharge.

Complications

Wound infection, abscess; intraabdominal abscess; suture line leak; duodenal fistula.

Follow-Up

Continue antiulcer medications for approximately 6 weeks. At that point, patients should be tested for *H. pylori* and treated if positive.

Northwestern Handbook of Surgical Procedures, edited by Richard H. Bell, Jr. and Dixon B. Kaufman. ©2005 Landes Bioscience.

Chapter 12

Major Hepatic Laceration: Open Repair

Michael A. West

Indications

A large proportion of liver injuries can be managed nonoperatively. The presence of a liver injury is not, by itself, an indication for operative exploration or repair. Operative control of liver hemorrhage is indicated with evidence of ongoing, hemodynamically significant bleeding.

Preop

Prior to performing exploratory laparatomy in a hemodynamically unstable patient, appropriate large-bore venous access must be present and resuscitation initiated. Blood for type and cross-match should be sent. A Foley catheter and nasogastric tube are placed prior to abdominal exploration. When performing exploratory laparotomy for trauma the surgeon should have adequate operative suction (two suctions), lighting, and the patient positioned so that the chest and/or mediastinum can be accessed intraoperatively. Antibiotic prophylaxis should be given prior to making the incision. A second-generation cephalosporin or other agents that cover aerobic and anaerobic enteric pathogens are generally used.

Step 1. Exploratory laparotomy (see Chapter 49) is performed.

Step 2. A retractor is inserted below the right costal margin to permit visual examination of the liver. Initially the liver is examined in situ, looking for evidence of deep lacerations and active bleeding. Gentle upward (anterior and cephalad) retraction may permit visualization of the undersurface of the liver. The condition of the gallbladder should also be carefully noted. It is not necessary to mobilize the liver in the absence of an actively bleeding injury because mobilization itself may precipitate bleeding. Clots should not be removed if there is no evidence of ongoing bleeding. If there is hepatic injury and ongoing hemorrhage, operative control may be required.

Step 3. Temporary control of active liver bleeding can often be accomplished using a combination of direct manual compression and packing. Laparotomy pads are packed between the liver and the diaphragm and between the liver and the retroperitoneum. Additionally, direct manual compression can be applied using two hands to compress the liver. If temporary control of bleeding is attained, direct pressure should be held for 10 minutes. Packing may be facilitated by partial mobilization of the liver (see below). After an appropriate interval, release pressure to determine if bleeding resumes. If not, cautiously remove the laparotomy packs. If bleeding resumes, replace the packs and restore hemostasis. Consider leaving the packs in place and returning for a "second look" laparotomy in 24-72 hours.

Northwestern Handbook of Surgical Procedures, edited by Richard H. Bell, Jr. and Dixon B. Kaufman. ©2005 Landes Bioscience.

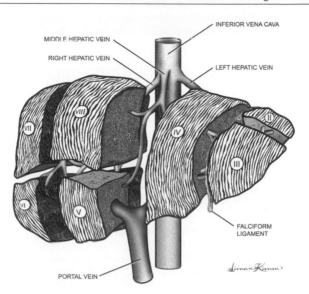

Figure 12.1. Liver segmental anatomy.

Step 4. If direct pressure and packing does not achieve hemostasis a Pringle maneuver should be performed. The Pringle maneuver involves using digital pressure or an atraumatic clamp across the portal triad to occlude the portal vein and hepatic artery. If this maneuver significantly slows hemorrhage it suggests that the injury within the parenchyma involves branches of these vessels. Several methods to identify and control such injuries are described below. More importantly, if the Pringle maneuver does not halt or significantly slow bleeding from the liver it suggests that hemorrhage is arising from the hepatic veins or retrohepatic vena cava.

Step 5. If bleeding continues after the Pringle maneuver, try repacking the patient. If bleeding is controlled or significantly slowed, consider closing the patient with the packs in place and returning in 24-72 hours to reassess and remove the packs. If repacking does not control the hemorrhage, an attempt at operative control is indicated. Operative measures to control hemorrhage frequently require complete mobilization of the liver. The surgeon should have long vascular instruments and a needle holder preloaded with a 5-0 vascular suture available when liver mobilization for bleeding is initiated.

Step 6. Hepatic mobilization is initiated by dividing the ligamentum teres and falciform ligament. This permits complete visualization of the anterior surface of the liver. Next, the lateral attachments are released by incising the triangular ligaments. Usually the left triangular ligament is divided first. The right triangular ligament is incised next. In many instances it is easiest to begin incising the right triangular ligament at the inferior edge. When fully mobilized the liver can be rotated (gently) to assess the posterior aspect of the right and left lobes. Mobilization is required to access the hepatic veins and retrohepatic vena cava. If the surgeon strongly suspects a major retrohepatic injury, consideration should be given to performing an atrial-caval shunt (see Step 12) prior to complete mobilization to avoid exsanguinating hemorrhage.

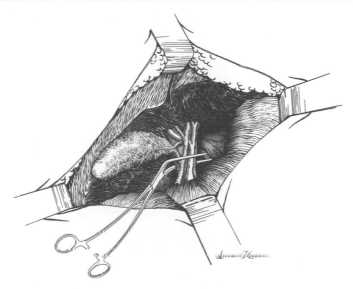

Figure 12.2. Pringle maneuver.

Step 7. Intrahepatic hemostasis can often be accomplished using the "finger-fracture" technique. This involves blunt division of the hepatic parenchyma in the direction of the laceration to expose bleeding intrahepatic vessels. Once identified, sites of bleeding can be controlled with clamps, ligatures, or surgical clips. Deep liver sutures may also be useful to control hemorrhage, but can produce significant ischemic damage to adjacent healthy parenchyma. Deep liver sutures (typically a 0-chromic suture on a round-tip needle) are placed in a horizontal mattress fashion on either side of the laceration. (Demonstration that manual compression controls hemorrhage should precede suture placement). Sutures are placed so as to achieve gentle compression of the interposed tissue. In some instances an absorbable mesh (e.g., polyglycolic acid) can be used to buttress the sutures.

Step 8. Electrocautery or the argon beam coagulator are frequently very useful for control of surface bleeding or small vessel bleeding from raw surfaces. When using conventional electrocautery, improved hemostasis is often achieved by turning the coagulation level to a very high level and positioning the cautery tip 2-3 mm above the bleeding surface. The resulting arc coagulates a wider superficial area. Care must be taken to remember to turn the cautery level back down before using the cautery on other structures. The gas jet of the argon beam coagulator permits the surgeon to more easily visualize the bleeding surgical field.

Step 9. In instances where multiple or deeper lacerations are found, bleeding can sometimes be controlled by wrapping the injured portion of the liver with polyglycolic acid mesh. The mesh should be cut, folded, and the edges sutured with a running absorbable suture so as to envelop a lobe and to achieve a slight degree of compression. However, the surgeon must be careful to avoid excess compression that may lead to infarction.

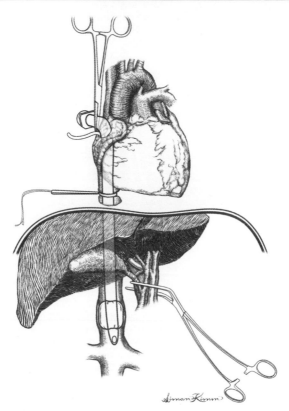

Figure 12.3. Atrial-caval shunt.

Step 10. An important adjunct to the hemostatic techniques described above is placement of a pedicle of omentum into the laceration. The omentum stimulates fibrin polymerization, provides a tissue seal for severed biliary radicles, minimizes dead space within the injured liver, and may hasten healing.

Step 11. If hemostasis has been achieved and if the patient is hemodynamically stable, nonviable liver tissue should be debrided to minimize late infectious complications. Debridement is not a priority and should not be performed during the initial "damage control" operation in hemodynamically unstable patients.

Step 12. Severe retrohepatic injuries have an extraordinarily high mortality irrespective of the surgical management employed. Prompt recognition of this lethal injury (continued venous bleeding with Pringle maneuver) is essential to achieve survival. Atrial-caval shunt permits vascular isolation of the liver and may permit the surgeon to visualize and repair the injury. A median sternotomy is performed and a 7-mm endotracheal (ET) tube (or a 36 F chest tube) is inserted into the inferior vena cava via the right atrial appendage. A hole must be carefully cut in the side of the tube (avoiding the balloon lumen of the ET tube) that will lie in the right atrium. An encircling tourniquet is passed around the vena cava cephalad to the injury (this is most easily achieved within the pericardial sac). A Pringle maneuver is performed along with inflation of the ET balloon in the infrahepatic cava (an additional encircling tourniquet is required below the liver if a chest tube is used) and clamping of the open end of the tube is performed to achieve vascular isolation. The injured vein is then repaired expeditiously to minimize warm liver ischemia.

Step 13. Hepatic resection should be reserved for the rare instances where the injury has already performed most of the resection or as a last resort to control uncontrollable exsanguinating hemorrhage.

Step 14. The role of liver drainage remains controversial. Drainage is not needed for most Grade I/II liver injuries. When drainage appears appropriate for Grade III-V injuries, one or more closed suction drains should be used. Open drainage techniques (e.g., Penrose drains) have a high incidence of septic complications and should be avoided.

Operative Principles

Complete abdominal exploration should be performed prior to attempting to control bleeding from liver injuries. Most liver injuries result in bleeding from relatively low pressure vessels (portal or hepatic veins). Therefore, packing and correction of coagulation abnormalities is frequently crucial to successful control of liver hemorrhage. Damage control techniques may be life-saving with severe hepatic injuries. A good understanding of hepatic vascular anatomy is important prior to attempting open repair of liver injury. Coagulopathy from hypothermia, hemodilution, massive blood transfusion, and coagulation factor depletion may contribute to ongoing hemorrhage. Angiographic techniques may be useful in selected unstable patients with extensive liver injuries or when operative techniques have failed.

Postop

Careful postoperative hemodynamic management and fluid management are indicated following repair of major hepatic injuries. Hypertension may aggravate bleeding and should be avoided. Hypothermia should be assiduously avoided. Coagulation parameters should be aggressively monitored and abnormalities corrected. Alterations in renal function are frequently encountered after severe hemorrhagic shock. Unstable patients should be monitored in an ICU, and strong consideration should be given to utilizing a pulmonary artery catheter to guide resuscitation, even in young, otherwise healthy patients. Marked elevations of transaminase levels suggest significant ischemic injury and may signal evidence of hepatic insufficiency.

Complications

Complications include rebleeding, hemobilia, and intrahepatic arteriovenous fistulae. Postoperative bile leaks are common and can often (but not always) be controlled with biliary stents or percutaneous drains. Infectious complications, such as subphrenic abscess or intrahepatic abscesses, are also frequent. Hepatic insufficiency, acute renal failure, ARDS, multiple organ dysfunction syndrome, and death may occur.

Follow-Up

The patient should be followed until all wounds have healed. Long-term follow-up depends on the nature of the underlying disease/injury.

Chapter 13

Hepaticojejunostomy: Roux-en-Y

Mark S. Talamonti

Indications

Roux-en-Y hepaticojejunostomy is indicated for reconstruction following resection for carcinoma of the proximal bile duct and hepatic duct bifurcation (Klatskin's tumors). When tumors of the distal third of the bile duct or head of the pancreas are not resectable, a bypass between the hepatic duct and a Roux-en-Y limb of jejunum may be done to relieve distal biliary obstruction. Benign indications for Roux-en-Y hepaticojejunostomy include: extrahepatic biliary stricture secondary to previous surgical injury, common bile duct obstruction secondary to recurrent stones, and distal common bile duct stricture secondary to chronic pancreatitis. Contraindications to resection and reconstruction for malignancy include distant metastases or extensive, bilateral liver involvement. Lymph node metastases outside the region of the porta hepatis are usually considered a contraindication to resection. Local extension of the tumor to include the main portal vein with thrombosis of the vein, and encasement of the main common hepatic artery are also contraindications to resection. Unilateral involvement of the proximal bile duct, portal vein, or hepatic artery branch may necessitate combined hepatic resection with biliary resection and reconstruction.

Preop

In cancer patients, a triphasic helical CT scan of the liver and porta hepatis is done for assessment of local tumor extension. Percutaneous transhepatic cholangiography is done to demonstrate the extent and location of tumors or strictures of the biliary system and for preoperative relief of biliary obstruction. Placement of the percutaneous stent through the hepatic duct stricture or tumor into the duodenum will facilitate identification and dissection at surgery. Hepatic artery angiography and portography may be required to delineate the relation of a tumor to the main vascular structures within the porta hepatis. On the morning of surgery, patients receive preoperative antibiotics and an epidural catheter for intraoperative and postoperative analgesia. Deep vein thrombosis (DVT) prophylaxis with subcutaneous heparin and/or sequential compression devices is used according to the patient's risk factors for thromboembolus.

Procedure

Step 1. The patient is placed in the supine position. General anesthesia is induced and the abdomen is prepared for exploration. An upper midline incision is suitable for most patients. If a right subcostal incision was used for a previous operation, the abdomen may be entered through the prior incision.

Northwestern Handbook of Surgical Procedures, edited by Richard H. Bell, Jr. and Dixon B. Kaufman. ©2005 Landes Bioscience.

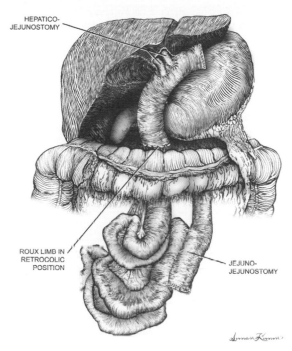

HEPATICO-
JEJUNOSTOMY

ROUX LIMB IN
RETROCOLIC
POSITION

JEJUNO-
JEJUNOSTOMY

13

Figure 13.1. Roux-en-Y hepaticojejunostomy.

Step 2. When operating for malignant strictures, exploration of the abdomen is done to rule out disseminated hepatic disease or peritoneal and omental metastases. In the absence of disseminated metastatic disease, exposure of the porta hepatis is begun.

Step 3. The falciform ligament is divided and the lateral suspensory ligaments of the liver are also divided to facilitate mobilization and retraction of the liver.

Step 4. A Kocher maneuver is done to facilitate mobilization and exposure of the distal common bile duct.

Step 5. Mechanical self-retaining retractors are placed in the upper abdomen, and the liver is retracted superiorly and laterally into the right upper quadrant. Manual countertraction is applied inferiorly and medially to provide exposure of the soft tissues in the porta hepatis

Step 6. If the gallbladder has not been previously removed then gentle traction is applied superiorly and anteriorly on the infundibulum of the gallbladder. The cystic duct is identified and dissected down to its junction with the common hepatic duct. The anterior peritoneum of the porta hepatis is completely divided across the bile duct and the hepatic artery. Circumferential dissection is then accomplished at the level of the cystic duct-common hepatic duct junction. A vessel loop is placed around the common bile duct, and the bile duct is then retracted laterally. The gallbladder is removed at this point or may be removed later en bloc with the specimen.

Step 7. If the gallbladder has been previously removed, identification of the bile duct is facilitated by the previously placed percutaneous biliary stents. By palpating the porta hepatis, the stent may be felt. Then the layer of peritoneum overlying the common hepatic duct can be divided. The peritoneum is divided medially to also expose the common hepatic artery. Once the common bile duct and the common hepatic artery have been identified, vessel loops are placed around these structures. The bile duct is retracted laterally and the hepatic artery retracted medially. The soft tissues between the common hepatic artery and the common bile duct are dissected toward the main bile duct. The portal vein is identified posteriorly and dissected free of the bile duct. In cases of malignant strictures, the relationship of the proximal bile duct and tumor to the portal vein and hepatic artery branches is inspected. If the tumor has not extended into the main portal vein or has not encased the common hepatic artery, the dissection is continued. When there is local extension, the usual bile duct tumor will involve the right hepatic artery and potentially the right portal vein. In these situations, consideration should be given to right hepatic lobectomy or trisegmentectomy with biliary reconstruction to the left hepatic duct. In the cases of a benign biliary stricture, care must also be taken to identify the right hepatic artery and the bifurcation of the common hepatic artery as these areas may be densely adherent to the area of stricture and inflammation.

Step 8. Once the anatomic relationship of the biliary tree to the vascular structures in the porta hepatis has been identified and cleanly dissected, preparation is begun for resection and reconstruction of the biliary duct. The mobilized distal bile duct is retracted superiorly and divided distally above the entrance of the bile duct into the head of the pancreas. The distal common duct is closed with a running 4-0 absorbable suture. The proximal end is retracted anteriorly and superiorly, and dissection of the posterior wall of the common bile duct off of the anterior wall of the portal vein is carried to the level of the hepatic duct bifurcation.

Step 9. In cases of benign biliary strictures, the common hepatic duct may be divided at the level of the right and left bifurcation if the ductal tissue is healthy at that point. It is at this point that the diameter of the common hepatic duct will be largest to facilitate subsequent biliary enteric anastomosis.

Step 10. In cases of malignant biliary obstruction, dissection into the hepatic parenchyma along the main right and left hepatic ducts may be required to obtain a clear proximal margin. In such cases bilateral hepaticojejunostomies may be required for reconstruction. Frozen-section examination of the proximal portions of the right and left ducts is done to determine if the tumor has been completely resected.

Step 11. In malignant cases, posterior lymph nodes in the porta hepatis are dissected completely off the portal vein with care taken to identify aberrant or anomalous right hepatic arteries. Dissection is carried proximally on the common hepatic artery back toward the level of the celiac trunk. The common hepatic artery lymph nodes are dissected circumferentially off of the hepatic artery. No such lymphadenectomy is required in cases of benign biliary strictures; however, the proximal biliary ducts are frequently surrounded by a dense inflammatory reaction with fibrosis. Thus, the dissection in benign cases may be relatively more difficult than in malignant situations.

Step 12. Reconstruction is begun with preparation of the Roux-en-Y limb. The proximal jejunum is divided with a linear stapling device approximately 15 cm distal to the ligament of Treitz. The mesentery is divided between clamps down to the level of the superior mesenteric artery.

Step 13. The distal end of the cut jejunum is brought through an opening in the transverse mesocolon to rest in a retrocolic position. If there is a large mass in the head of the pancreas or dissection is particularly difficult in this area, the limb may be put in an antecolic position. A primary biliary-to-enteric anastomosis between the end of the common hepatic duct and the antimesenteric border of the jejunum is done in a single layer with interrupted 5-0 absorbable sutures.

Step 14. Because of the concern for possible postoperative stricture formation, the end of the Roux-en-Y limb may be brought to the anterior abdominal wall and a 7 mm fenestrated stent placed across the anastomosis. This stent is then brought out through the top of the Roux-en-Y limb and through the anterior abdominal wall. This technique provides a stent across the anastomosis and internal drainage of the biliary tract. In addition, postoperative cholangiography can be performed through this stent to assess for postoperative stricture formation or anastomotic leak.

Step 15. The jejunum is anchored to the anterior abdominal wall with interrupted 3-0 absorbable sutures at the site of entrance of the jejunal catheter stent. Gastrointestinal tract continuity is then reestablished by stapling the end of the proximal jejunum to the descending limb of the Roux-en-Y segment with a linear stapling device. The enterotomies used to introduce the stapling device are closed with a non-cutting stapler. Two abdominal drains are placed in the right upper quadrant, posterior and anterior to the biliary anastomosis. The ends of these drains are brought out through stab wounds in the right upper quadrant, and the abdomen is closed in layers and the skin closed with staples.

Postop

Nasogastric decompression is maintained until bowel function returns. Drain output is monitored for bile. When the drain output is clear and less than 30 cc per day for 2-3 days in a row, the drains are removed. Antibiotics are discontinued after the 24-hour perioperative period.

Complications

Early complications include trauma to the portal vein or hepatic artery during dissection of the porta hepatis and postoperative hemorrhage. Subsequent complications also include anastomotic leakage with persistent biliary fistula, subhepatic abscess formation, and small bowel obstruction secondary to angulation or torsion of the small bowel anastomosis. The most important long-term complication is anastomotic stricture secondary to ischemia or tumor recurrence.

Follow-Up

If employed, the jejunal stent may be removed after a few weeks if the patient is well and a postoperative cholangiogram shows a patent anastomosis without narrowing. For patients with malignant disease, liver function tests and clinical status should be followed at regular intervals. A change in status or chemistries requires reevaluation with imaging to rule out tumor recurrence. For patients operated for benign strictures, regular lifelong surveillance of liver chemistries is indicated because of a significant rate of recurrent stricture formation, which may be asymptomatic but lead to secondary biliary cirrhosis if not corrected.

Chapter 14

Cholecystectomy with Cholangiography: Open

Jay B. Prystowsky

Indications

The indications for open cholecystectomy are the same as for laparoscopic cholecystectomy AND inability to perform laparoscopic cholecystectomy (which, in general, is the procedure of choice). Indications for cholecystectomy include symptomatic cholelithiasis (acute or chronic cholecystitis), gallstone pancreatitis, acalculous cholecystitis, or choledocholithiasis.

Preop

Antibiotics are administered in cases of acute disease, choledocholithiasis, or age >65 years.

Procedure

Step 1. A right subcostal incision is performed.

Step 2. The costal margin is retracted cephalad; the hepatic flexure of the colon and the duodenum are retracted inferiorly.

Step 3. Grasping the fundus of the gallbladder with a clamp, it is lifted anteriorly and away from the liver.

Step 4. The peritoneum overlying the gallbladder is incised with cautery within a few millimeters of the liver.

Step 5. Progressively retracting it away from the liver, the gallbladder is dissected from Glisson's capsule in the gallbladder fossa, moving downward towards the porta hepatis. It is important to dissect close to the wall of the gallbladder.

Step 6. The cystic artery and cystic duct are identified.

Step 7. The cystic duct is dissected down to its junction with the common duct.

Step 8. The common duct immediately proximal and distal to the entrance of the cystic duct is identified to verify anatomy.

Steps 9-15 describe intraoperative cholangiography, which may be performed in selected cases. Indications for cholangiography generally include: elevated liver enzymes, stone in common bile duct either documented preoperatively or discovered by palpation intraoperatively, dilated common bile duct, recent gallstone pancreatitis, or difficulty dissecting or identifying biliary anatomy.

Step 9. To prepare for cholangiography, a ligature is placed proximally at the junction of the cystic duct and gallbladder.

Step 10. A small opening is made in the cystic duct and a cholangiocatheter (4-5 F) is passed into the duct for about 1-2 cm.

Step 11. The catheter is secured with a ligature or clip. Two 30 cc syringes are attached to the catheter with a three-way stopcock and extension tubing. One is filled with saline, the other with contrast diluted 50%. Saline is injected to confirm

Northwestern Handbook of Surgical Procedures, edited by Richard H. Bell, Jr. and Dixon B. Kaufman. ©2005 Landes Bioscience.

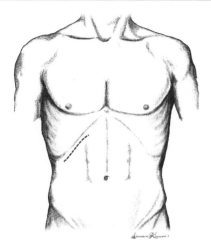

Figure 14.1. Cholecystectomy, open. Incision.

14

there are no leaks at the site of catheter entrance into the cystic duct. It should be possible to aspirate bile if the catheter is properly positioned. Before injecting dye, air bubbles should be eliminated from the catheter and tubing.

Step 12. The patient is then placed in the Trendelenburg position and tilted to the right (to bring the common duct "off" the spinal column).

Step 13. Contrast is injected under fluoroscopic guidance.

Step 14. Easy flow of contrast distally into the duodenum and proximally into the right and left biliary radicals along with absence of filling defects constitutes a normal exam.

Step 15. The catheter is withdrawn and the cystic duct is ligated distal to the catheter entrance site. The cystic duct may then be transected.

Step 16. The cystic artery is ligated with nonabsorbable suture and transected between ligatures. The gallbladder is removed.

Step 17. The abdominal wall is closed in layers.

Postop

Diet may usually be instituted within 24 hours. Parenteral narcotics for pain are switched to oral prior to discharge.

Complications

Major complications include injury to the common bile duct and bile leak from the cystic duct stump; other surgical complications include wound infection and postoperative bleeding.

Follow-Up

Patients should be seen at 1-2 weeks and again at approximately 6 weeks. Most patients experience excellent relief of pain; 5% of patients will continue to have discomfort as they experienced preoperatively (post-cholecystectomy syndrome).

14

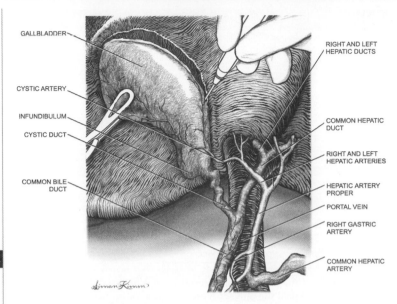

Figure 14.2. Open cholecystectomy.

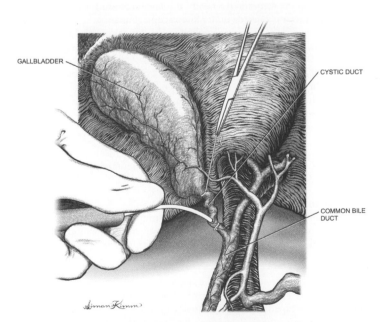

Figure 14.3. Open cholangiogram.

Chapter 15

Cholecystectomy with Cholangiogram: Laparoscopic

Kenric M. Murayama

Indications

The indications for laparoscopic cholecystectomy include symptomatic gallstone disease (chronic cholecystitis or acute cholecystitis) or acute acalculous cholecystitis. Cholangiography may be done in selective cases. In general, the indications for cholangiography include choledocholithiasis, dilated common bile duct, recent gallstone pancreatitis without preoperative ERCP, or confusion about the anatomical orientation intraoperatively.

Preop

A first- or second-generation cephalosporin or an antibiotic of equivalent coverage is given 30 minutes prior to surgery. A nasogastric or orogastric tube is placed to decompress the stomach.

Procedure

Step 1. The entire abdomen is prepped and draped in standard sterile fashion. Access is gained by either the Veress needle technique (closed technique) or open technique. If a needle is used, a drop of saline is placed in the needle after insertion and should move downward with respiration if the needle is in the peritoneal cavity. If the needle is properly positioned, CO_2 insufflation is initiated to a pressure of 15 mm Hg.

Step 2. The first trocar is placed at or in proximity to the umbilicus. If a 10 mm telescope is used, a 10 mm trocar is placed in this location and if a 5 mm telescope is used, then a 5 mm trocar can be placed.

Step 3. The patient is placed in reverse Trendelenburg position and the other trocars are placed under direct visualization. A 10 mm trocar is placed in the subxiphoid epigastric region; a 5 mm trocar is placed in the right, subcostal, midclavicular line; and a 5 mm trocar is placed in the right subcostal, anterior axillary line location.

Step 4. If the patient has had acute cholecystitis, there may be adhesions to the gallbladder. The duodenum and/or colon may be adherent to the surface of the gallbladder. Therefore, while electrosurgical cautery can generally be used to facilitate the dissection of adhesions, cautery should be avoided if the duodenum or hepatic flexure of the colon is in proximity.

Step 5. The fundus of the gallbladder is grasped with an instrument placed through the right subcostal, anterior axillary line port and the tip of the gallbladder is retracted cephalad. The infundibulum of the gallbladder is retracted caudad and to

Northwestern Handbook of Surgical Procedures, edited by Richard H. Bell, Jr. and Dixon B. Kaufman. ©2005 Landes Bioscience.

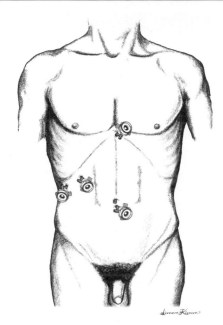

Figure 15.1. Laparoscopic cholecystectomy. Port placement.

the patient's right with a second grasper that is placed through the right subcostal midclavicular line port.

Step 6. A Maryland dissector is used to clear the peritoneum over the infundibulum and cystic duct. The cystic duct must be clearly identified as it exits the infundibulum of the gallbladder and traverses toward the common bile duct. Once the peritoneum overlying the cystic duct is opened, the cystic duct is cleared of its adventitial attachments circumferentially. Retracting the infundibulum toward the patient's left facilitates dissection of the lateral side of the cystic duct. The infundibulum is again retracted to the patient's right and the triangle of Calot is entered, hugging the edge of the cystic duct. An adequate segment of cystic duct is cleared.

Step 7. Prior to division of the cystic duct, a decision must be made regarding the need for a cholangiogram. If a cholangiogram is to be performed, a clip is placed across the cystic duct near the infundibulum. A transverse opening is created in the cystic duct. A "flash" of bile confirms that the opening is in the cystic duct. A cholangiocatheter is placed into the cystic duct and threaded distally toward the common bile duct. The catheter options include a balloon or straight catheter, and either a cholangiocatheter clamp or clips can be used to secure the catheter.

Two 30 cc syringes are attached to the catheter with a three-way stopcock and extension tubing. One is filled with saline, the other with contrast diluted 50%. Saline is injected to confirm there are no leaks at the site of catheter entrance into the cystic duct. It should be possible to aspirate bile if the catheter is properly positioned. Before injecting dye, air bubbles should be eliminated from the catheter and any extension tubing. Fluoroscopy or multiple static films can be used to verify the

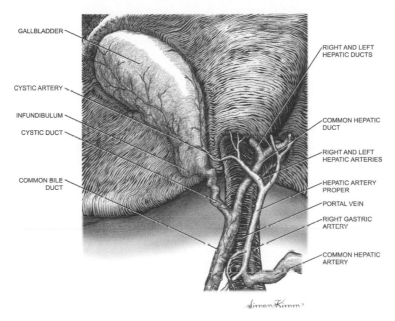

GALLBLADDER

RIGHT AND LEFT
HEPATIC DUCTS

CYSTIC ARTERY

INFUNDIBULUM

CYSTIC DUCT

COMMON HEPATIC
DUCT

RIGHT AND LEFT
HEPATIC ARTERIES

COMMON BILE
DUCT

HEPATIC ARTERY
PROPER

PORTAL VEIN

RIGHT GASTRIC
ARTERY

COMMON HEPATIC
ARTERY

15

Figure 15.2.Laparoscopic cholecystectomy. Anatomy.

presence or absence of common bile duct stones. If there are no stones, the operation can proceed. If there are common duct stones, there are four options:

1. complete cholecystectomy and have a postoperative ERCP performed;
2. complete cholecystectomy and have intraoperative ERCP performed;
3. laparoscopic common bile duct exploration; or
4. convert to open common bile duct exploration.

The decision depends on availability of a skilled endoscopist and the surgeon's experience with laparoscopic common bile duct exploration (see Chapter 17).

If the cholangiogram has been completed and a common bile duct exploration is not necessary or has also been completed, the cystic duct is divided between clips (ordinarily, two clips are placed on each side).

Step 8. The cystic artery is identified and cleared of surrounding attachments. Special care is taken to ensure that the artery is not the right hepatic artery by following it and observing its termination in the gallbladder. When this has been verified, the cystic artery is divided in continuity between clips (ordinarily, two clips are placed on each side).

Step 9. The infundibulum of the gallbladder is retracted anteriorly and cephalad progressively allowing the gallbladder to be dissected from the liver bed using cautery and blunt or sharp dissection.

Step 10. Once amputated the gallbladder is placed into a specimen retrieval bag and removed through either the subxiphoid or the periumbilical port. If the gallbladder is exceedingly large, full of gallstones, or contains large stones, it may not be possible to safely remove the gallbladder. Options include crushing the stones inside the gallbladder with a clamp, removing many stones/stone fragments to help decompress the gallbladder, and/or enlarging the port incision.

Step 11. Once the gallbladder is removed, the liver bed is examined to be sure there is no bleeding or bile leakage.

Step 12. The 10 mm incisions are closed at the fascial level. All skin incisions are closed with absorbable subcuticular sutures.

Addsteps. Conversion to open should occur if the anatomy is unclear, there is excessive bleeding, or if a complication such as common duct injury occurs.

Postop

Patients are started on clear liquids on the evening of surgery and may have their diet advance *ad libitum*. Patients are either sent home the day of surgery or in 23 hours.

Complications

Major complications include bleeding, common duct injury, leakage of bile from the cystic duct stump, duodenal injury, or other bowel injury.

Follow-Up

The patient should be seen in 1-2 weeks to examine wounds and be seen later by either the surgeon or referring physician to confirm resolution of preoperative symptoms.

15

Chapter 16

Common Bile Duct Exploration: Open

Jay B. Prystowsky

Indications

In general, open common duct exploration is indicated when stones are discovered by cholangiography during open cholecystectomy. It may be indicated when stones are discovered during laparoscopic cholecystectomy and the surgeon is not familiar with the technique of laparoscopic duct exploration. Palpable stones in the common bile duct at the time of open cholecystectomy are another indication. An alternative therapy for stones in the common bile duct is postoperative endoscopic extraction via ERCP. Common duct exploration should be strongly considered when stones are large or multiple or there are anatomic considerations that would make the stones not amenable to endoscopic extraction.

Preop

Antibiotic prophylaxis is indicated. The early steps of the operation are described under open cholecystectomy with cholangiography.

Procedure

Step 1. Once the common duct has been identified, its anterior wall should be exposed for about 2.5-3 cm; care should be taken to avoid dissection along its lateral walls since that is where its blood supply exists.

Step 2. A #15 blade is used to create a small rent in the anterior wall of the duct, and Potts scissors are used to enlarge the rent in a longitudinal fashion for about 2 cm; stay sutures are placed on either side of the common bile duct incision to keep the aperture open.

Step 3. Randall stone forceps are passed distally and then proximally to clear the duct of stones by directly grasping them.

Step 4. A choledochoscope is useful to identify residual stones and assist in their extraction.

Step 5. An appropriately sized T-tube is placed into the common duct, and the common duct closed over the tube with a series of interrupted 4-0 absorbable sutures.

Step 6. A cholangiogram is performed to ascertain that the duct is clear of stones.

Step 7. A drain is placed near the common bile duct opening and brought out through a separate stab incision.

The remainder of the case proceeds as for open cholecystectomy.

Northwestern Handbook of Surgical Procedures, edited by Richard H. Bell, Jr. and Dixon B. Kaufman. ©2005 Landes Bioscience.

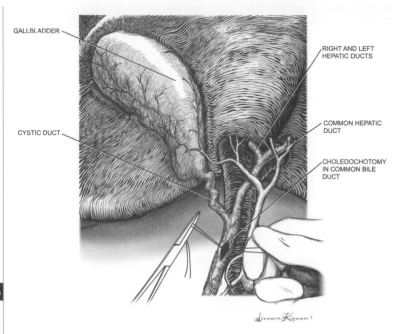

Figure 16.1. Common bile duct exploration.

Postop

The peritoneal drain can be removed in 24-48 hours if there is no bile leakage. The T-tube is initially placed to gravity drainage. Before discharge, a cholangiogram through the tube should be performed. If negative, the tube can be capped and the patient discharged.

Complications

Complications related to the T-tube are predominantly dislodgement or kinking. Retained stones may be present on follow-up cholangiogram and may require removal endoscopically or through the T-tube tract.

Follow-Up

About 2 weeks after surgery, the T-tube may be removed.

Chapter 17

Common Bile Duct Exploration: Laparoscopic

Kenric M. Murayama

Indications

Laparoscopic common bile duct exploration is indicated for the presence of common bile duct stone(s) during laparoscopic cholecystectomy. Usually stones are detected after an intraoperative cholangiogram (see Chapter 15).

Preop

Preoperative preparation is the same as for laparoscopic cholecystectomy with cholangiogram. The patient should be on a fluoroscopy-capable operating room table. A choledochoscope should be available. The initial steps of the operation are described in the chapter on laparoscopic cholecystectomy with cholangiogram. Two options for laparoscopic common bile duct exploration are possible: transcystic duct exploration of the common bile duct or choledochotomy (similar to open common bile duct exploration).

Procedure

Laparoscopic Transcystic Duct Exploration of the Common Bile Duct

Step 1. After the intraoperative cholangiogram reveals presence of common bile duct stones, transcystic duct exploration can be undertaken via the same hole in the cystic duct created for the cholangiogram. However, a larger hole with dilation of the cystic duct may be necessary to remove stones.

Step 2. A balloon catheter setup is utilized to dilate the cystic duct. The cystic duct should be gradually dilated over a period of 3-5 minutes. The cystic duct should never be dilated to a diameter larger than the common bile duct. The cystic duct needs to be dilated so that it is at least as large as the largest stone to be removed.

Step 3. Choledochoscopy can be performed through the cystic duct incision to visualize and localize the common duct stone(s). A choledochoscope is used that has a working channel of at least 1.2 mm. Body-temperature saline is used to irrigate the common bile duct to aid in visualization. If a stone is encountered, it should be removed before looking for more stones since failure to do so can result in stones first visualized floating up into the proximal bile ducts in the liver.

Step 4. To remove stones, a straight #4 wire basket (2.4 F) is preferable and should be threaded through the working channel of the choledochoscope. The wire basket is passed beyond the stone and opened. The stone is entrapped when the basket is withdrawn. Once entrapped, the stone should be gently grasped and the basket pulled snugly up against the end of the choledochoscope. Both the basket and choledochoscope are withdrawn completely as a unit. This process is repeated until all stones are completely removed.

Northwestern Handbook of Surgical Procedures, edited by Richard H. Bell, Jr. and Dixon B. Kaufman. ©2005 Landes Bioscience.

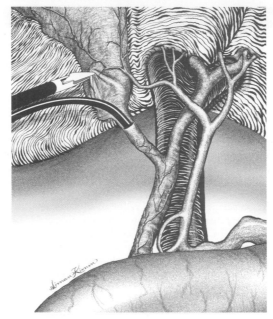

Figure 17.1. Laparoscopic transcystic common bile duct exploration.

Step 5. A completion cholangiogram is performed by reinserting a catheter through the cystic duct and securing it in place so that there is no leakage of contrast.

Step 6. A cystic duct tube for drainage can be inserted if there is concern for retained stones.

Laparoscopic Choledochotomy

This procedure is performed much as an open common duct exploration is performed. It requires the surgeon to have the capability to perform intracorporeal suturing and knot-tying. It is best to perform the common duct exploration before removal of the gallbladder since the clamps on the gallbladder can be used to retract the liver and to place traction on the common bile duct.

Step 1. Side-by-side stay sutures of 5-0 monofilament are placed about 2 mm apart in the wall of the common bile duct, just below the cystic duct-common bile duct junction. A longitudinal choledochotomy approximately 1 cm in length is created using microdissection laparoscopic shears. Any bile leakage is aspirated.

Step 2. The common bile duct is irrigated with body-temperature sterile saline to try to "float" any gallstones out via the choledochotomy.

Step 3. The choledochoscope is placed through the choledochotomy and the method for stone retrieval/removal is similar to that described above.

Step 9. Once all stones have been removed, a T-tube of appropriate size is fashioned and passed into the abdominal cavity. The T-tube is placed into the choledochotomy and the common duct closed around the tube with a series of interrupted 4-0 absorbable sutures. In either procedure, a suction drain is placed to monitor for leakage of bile from the cystic duct closure or the choledochotomy.

17

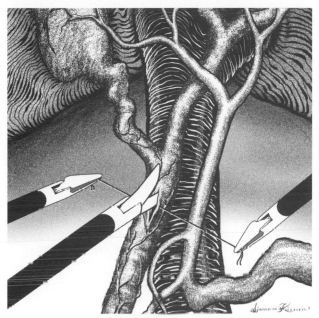

Figure 17.2. Laparoscopic choledochotomy.

Postop

In general, care is similar to that described for laparoscopic cholecystectomy. The peritoneal drain can be removed in 24-48 hours if there is no evidence of bile leakage. The T-tube or cystic duct drain is left in place for approximately 2 weeks before obtaining a tube cholangiogram. If there are no retained stones, the tube can be removed in the outpatient office.

Complications

Retained common duct stones may require endoscopic removal. Injury to the common bile duct may occur during common bile duct exploration if it is not carefully done. Bile duct stricture can be a long-term complication.

Follow-Up

Patients should be followed short-term at intervals until tubes are removed and liver function tests are normal. Long-term follow-up is described under laparoscopic cholecystectomy.

17

Chapter 18

Transduodenal Sphincteroplasty

Woody Denham

Indications

Transduodenal sphincteroplasty is indicated for the treatment of sphincter stenosis or dysfunction in selected cases. It is also indicated in the presence of multiple common duct stones in a nondilated system or to remove an impacted stone at the ampulla of Vater that is not removable by any other means.

Preop

A perioperative prophylactic antibiotic is administered 30 minutes prior to incision. Deep venous thrombosis prophylaxis with sequential compression devices or subcutaneous heparin is provided dependent on patient risk factors.

Procedure

Step 1. The abdomen is entered through a right subcostal incision. A Kocher maneuver of the duodenum is performed.

Step 2. The common bile duct is exposed above the duodenum. Access to the common duct is required to aid in the performance of a sphincteroplasty. If the sphincteroplasty is being done in combination with a common bile duct exploration, access is through the choledochostomy. If the common duct does not need to be opened, access can be gained through the cystic duct stump.

Step 3. A #4 or #5 Fogarty catheter is passed through the cystic duct or choledochostomy into the duodenum. The balloon is inflated once the catheter tip is in the duodenum and the catheter pulled back snugly so that the balloon impacts against the ampulla.

Step 4. A longitudinal incision is made in the duodenum over the ampulla. The incision is centered over the Fogarty balloon on the opposite wall of the duodenum.

Step 5. The ampulla of Vater is located and 3-0 silk stay sutures placed on either side of the ampulla to elevate it.

Step 6. The stay sutures are lifted forward, the balloon deflated, and the catheter partially withdrawn into the bile duct. A metal probe or grooved director is placed in the bile duct. The bile duct will be at the 11 o'clock position on the ampulla. The pancreatic duct is in the 4 o'clock position.

Step 7. With cautery, the ampulla is progressively incised over the probe, unroofing the bile duct 3-4 mm at a time. Each time a new 3-4 mm incision is made, interrupted 4-0 absorbable sutures should be placed from the bile duct mucosa to the duodenal mucosa on each side of the opening, being careful not to compromise the pancreatic duct opening.

Northwestern Handbook of Surgical Procedures, edited by Richard H. Bell, Jr. and Dixon B. Kaufman. ©2005 Landes Bioscience.

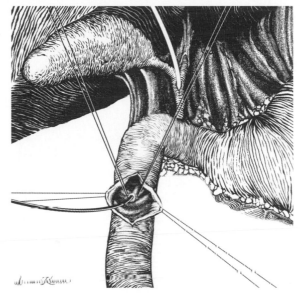

Figure 18.1. Transduodenal sphincteroplasty. Ampulla of Vater exposed and cannulated from the cystic duct.

Step 8. The opening in the bile duct should be extended until a Bakes dilator equal to the diameter of the common duct passes easily.

Step 9. A final 4-0 absorbable suture is placed at the apex.

Step 10. A pancreatic duct sphincteroplasty can be performed in the same manner except that the probe is placed in the pancreatic duct (4 o'clock position). If sphincteroplasties of the common bile duct and pancreatic duct are performed, the mucosa of the common wall between the two is sutured together with interrupted absorbable suture. The other wall of each duct is attached to the duodenal mucosa as previously described.

Step 11. The duodenum is closed transversely in two layers to prevent narrowing.

Postop

Intravenous fluids and nothing by mouth should be continued until bowel function returns. Antibiotics are discontinued after 24-hours, except in patients with preexisting cholangitis. Pain is controlled by intravenous narcotics until the patient is tolerating a diet.

Complications

Wound infection is a risk in the face of infected bile. Bleeding from the sphincteroplasty may occur. If the sphincteroplasty is carried too deep, the duodenal wall may be perforated and retroperitoneal leakage occur. The duodenal closure can leak.

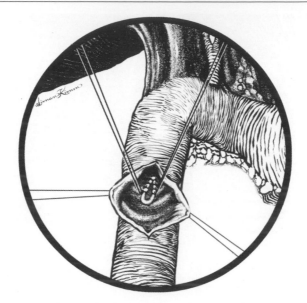

Figure 18.2. Transduodenal sphincteroplasty of ampulla of Vater complete.

18

Follow-Up

Patients should be followed until wounds are healed. If a T-tube was left in the common duct as part of the operation, a T-tube cholangiogram should be done as an outpatient and the T-tube removed about 2 weeks postoperatively.

Chapter 19

Pancreatic Necrosis: Debridement

Woody Denham

Indications

Debridement is indicated for infected pancreatic necrosis following an attack of acute pancreatitis.

Preop

The patient should be typed and cross-matched for 2-4 units of blood because debridment of the retroperioneum may be associated with moderate or extensive blood loss. Patients should be continued on preoperative antibiotics.

Procedure

Step 1. After prepping and draping the entire abdomen, a midline or bilateral subcostal incision is made

Step 2. The pancreas is usually approached by entering the lesser sac through the gastrocolic ligament. The gastrocolic omentum is ordinarily opened just below the gastroepiploic vessels. If entry into the lesser sac is difficult due to the inflammation, it may be safer to enter the lesser sac immediately adjacent to the stomach to avoid injury to the transverse mesocolon and middle colic vessels.

Step 3. If the retroperitoneal inflammation prevents entry through the gastrocolic omentum, the pancreas may be approached through the base of the transverse mesocolon. An incision to the right of the middle colic vessels is used to debride the head of the gland while an incision to the left of the vessels is used to debride the body and tail.

Step 4. The necrotic portions of the pancreas and peripancreatic tissue are removed manually. Laparotomy sponges may be useful. In addtion, ring forceps are helpful to remove nonviable tissue. Sharp dissection should be used very sparingly.

Step 5. Bleeding is controlled with direct pressure or suture ligatures.

Step 6. At least two drains are left in the pancreatic bed. These should be placed over the head and the body and tail of the gland. They are brought through separate incisions in the abdominal wall.

Step 7. A feeding jejunostomy tube and gastrostomy tube for gastric decompression should be placed.

Step 8. The abdomen is closed. Retention sutures should be considered.

Step 9. If the patient has extensive necrotic tissue in the retroperitoneum which cannot be debrided in one operation, the lesser sac can be marsupialized and packed with gauze and the abdominal wall left open, with the cavity dressings changed in the intensive care unit under sedation. Alternatively, the patient can be returned to the operating room several days in a row until the nonviable tissue has all been removed. In this case, temporary closure of the abdomen with Gore-Tex® or silicone sheeting sewn to the fascia around the periphery of the wound can be quite helpful.

Northwestern Handbook of Surgical Procedures, edited by Richard H. Bell, Jr. and Dixon B. Kaufman. ©2005 Landes Bioscience.

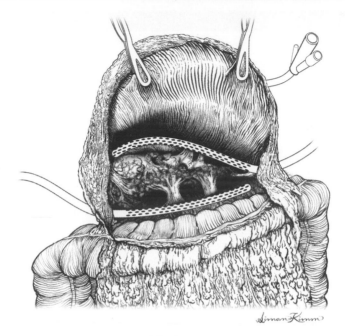

Figure 19.1. Pancreatic necrosis debridement.

When returning to the operating room, the sheeting can be reopened in its midportion and reclosed without the necessity of disturbing the fascia.

Postop
The patient is often critically ill and requires intensive care. The drains are placed on low suction, and they are irrigated each shift to prevent obstruction. The gastrostomy tube is placed on gravity drainage. Jejunostomy feeds can be started within 24-48 hours.

Complications
One of the most common causes of death in patients who have undergone pancreatic debridement is persistent or recurrent retroperitoneal sepsis. If the patient is not improving postoperatively, a CT scan should be performed to determine if undrained fluid collections are present or if additional nonviable tissue is present. Complications of pancreatic debdridement include respiratory or renal failure, retroperitoneal hemorrhage, colonic ischemia, and gastrointestinal fistulae.

Follow-Up
Patients who survive the acute episode of necrosis will need to be followed for the development of pancreatic endocrine and exocrine insufficiency. Patients who are treated by the open packing method will usually develop a ventral hernia which ultimately requires repair.

Chapter 20

Pancreaticoduodenectomy: Whipple Procedure

Mark S. Talamonti

Indications

The majority of patients undergo pancreaticoduodenectomy for ductal adenocarcinoma of the pancreatic head. Other primary tumor types for which pancreaticoduodenectomy is indicated include carcinoma of the ampulla, duodenal carcinoma, carcinoma of the distal common bile duct, islet cell carcinoma, and mucinous cystic neoplasms. Absolute contraindications include distant metastases to the liver, peritoneum, omentum, or lungs. Lymph node metastases outside the region of the head of the pancreas are usually considered a contraindication to resection. Other contraindications include portal/superior mesenteric vein thrombosis and encasement of the superior mesenteric or hepatic artery.

Preop

Preoperative radiologic staging is done by triphasic helical CT scan of the pancreas with 1-2 mm sections through the head of the gland. In patients with profound jaundice (bilirubin greater than 20 mg/dl) and malnutrition, consideration should be given to placement of an endoscopic biliary stent with a period of nutritional supplementation prior to the procedure. An oral antibiotic and mechanical bowel prep is indicated since segmental resection of the colon is occasionally required in conjunction with tumor resection. On the morning of surgery, patients receive a preoperative prophylactic antibiotic and an epidural catheter for intraoperative and postoperative analgesia. Deep vein thrombosis prophylaxis with subcutaneous heparin and sequential compression boots should be employed.

Procedure

Step 1. The patient is placed in the supine position, general anesthetic is induced, and the abdomen is prepared for exploration. A relatively thin patient may be explored through a midline incision. For patients with previous mid or lower abdominal surgery or with a wide costal margin, a bilateral subcostal incision is employed.

Step 2. Exploration of the entire abdomen is done to rule out hepatic, peritoneal, or omental metastases. In the absence of disseminated metastatic disease, exposure of the pancreas is begun.

Step 3. The transverse mesocolon is retracted inferiorly and the omentum and the stomach are retracted superiorly and anteriorly. The lesser sac is entered through the greater omentum. The middle colic vein is identified on the posterior surface of

Northwestern Handbook of Surgical Procedures, edited by Richard H. Bell, Jr. and Dixon B. Kaufman. ©2005 Landes Bioscience.

the transverse mesocolon and dissected down to its junction with the superior mesenteric vein.

Step 4. The inferior neck of the pancreas is inspected for evidence of tumor extension from the inferior edge of the pancreas on to the superior and anterior surfaces of the superior mesenteric vein and into the root of the small bowel mesentery. In the absence of these findings, dissection is continued.

Step 5. The third portion of the duodenum and the uncinate process are retracted superiorly and the hepatic flexure and transverse mesocolon are retracted inferiorly. Attachments between these organs are divided with a combination of sharp dissection and electrocautery. Inspection is carried out for evidence of locally advanced extension into the root of the small bowel or into the retroperitoneum.

Step 6. An extended Kocher maneuver is then performed, mobilizing the duodenum and the head of the pancreas off of the retroperitoneal structures. The inferior vena cava is identified, and the posterior pancreatic soft tissues including the posterior pancreatic lymph nodes are completely dissected off of the inferior vena cava and the aorta. Extension of the Kocher maneuver up to the porta hepatis is then performed. En bloc dissection of the retroperitoneal soft tissues is accomplished up to the level of the left renal vein.

Step 7. The gallbladder is retracted superiorly and anteriorly. The cystic duct is identified and dissected down to its junction with the common hepatic duct. The anterior peritoneum of the porta hepatis is completely divided across the bile duct and the hepatic artery. Circumferential dissection is then accomplished at the level of the cystic duct and common hepatic duct confluence. A vessel loop is placed around the common bile duct, and the bile duct is retracted laterally.

Step 8. The common hepatic artery is identified and dissected circumferentially to expose the take-off of the gastroduodenal artery. The right gastric artery is identified, ligated, and divided if there is no plan to perform a pylorus-preserving procedure. The right gastric artery should be preserved if a pylorus-preserving pancreaticoduodenectomy is planned. Dissection is then carried proximally on the common hepatic artery back toward the level of the celiac trunk. The common hepatic artery lymph nodes are dissected circumferentially off of the hepatic artery and sent for frozen section. If biopsies at the level of the hepatic artery and celiac trunk fail to reveal evidence of metastatic cancer, the procedure is continued.

Step 9. The gastroduodenal artery is identified and ligated just distal to its origin from the common hepatic artery with care taken to avoid narrowing of the common hepatic artery. The gastroduodenal artery is oversewn with a 4-0 polypropylene suture ligature.

Step 10. A vessel loop is placed around the common hepatic artery and traction is applied medially. The vessel loop around the common bile duct is retracted laterally and the soft tissues between the common hepatic artery and the common bile duct are then dissected toward the main specimen. The anterior surface of the portal vein is identified and dissected down to the superior neck of the pancreas. This area is inspected to determine if there is tumor extension onto the portal vein. In the absence of any locally advanced extension into the common hepatic artery, or circumferentially around the portal vein, the dissection is continued.

Step 11. The posterior lymph nodes in the porta hepatis are then dissected completely off of the portal vein. Care is taken to identify any aberrant right hepatic arteries, which generally lie posterolateral to the vein.

20

Step 12. The gallbladder is dissected off of the inferior surface of the liver, and the cystic artery originating from the right hepatic artery is identified, ligated, and divided. The common hepatic duct is divided approximately 1 cm distal to the hepatic duct bifurcation and proximal to the cystic duct junction.

Step 13. If a pylorus-preserving Whipple procedure is being done, the subpyloric lymph nodes are dissected off of the inferior portion of the duodenum and the duodenum is divided approximately 2-3 cm distal to the pyloric valve with a 60 mm linear stapling device. In the standard Whipple procedure, which includes resection of the distal stomach, the greater omentum is dissected off of the transverse mesocolon up to the level of the confluence of the left gastroepiploic and right gastroepiploic arteries on the greater curvature of the stomach. The bare area between this confluence is chosen as the line of division on the greater curvature. On the lesser curvature, the branches of the left gastric artery are identified and at approximately the level of the second or third branch, the lesser omentum is dissected off the lesser curvature of the stomach, and then the stomach is divided using an 80 mm linear stapling device. The proximal portion of the staple line is oversewn with a running 3-0 polypropylene Lembert suture. The distal stomach is retracted to the right of the porta hepatis.

Step 14. The proximal jejunum is then divided approximately 10-15 cm distal to the ligament of Treitz with a 60 mm linear stapling device. Mesenteric branches to the proximal jejunum and distal duodenum are divided between clamps and oversewn on the mesenteric side with 3-0 absorbable suture ligatures to assure complete hemostatic control of the mesenteric vessels. Dissection is carried back to the level of the superior mesenteric artery and uncinate process.

Step 15. The fourth portion of the duodenum and the first portion of the jejunum are delivered through the retroperitoneum from left to right beneath the mesenteric vessels.

Step 16. In preparation for the division of the neck of the pancreas, four separate 3-0 polypropylene figure-of-eight stitches are placed on either side of the inferior and superior neck of the pancreas to control the inferior and superior pancreatic arterial arcade.

Step 17. The neck of the pancreas is carefully dissected off the anterior surface of the superior mesenteric vein and the anterior surface of the portal vein. The pancreas is divided with electrocautery. Care is taken to identify the pancreatic duct as it is divided.

Step 18. The pancreatic neck is dissected off the right lateral surface of the portal vein. The superior uncinate vein and the inferior uncinate vein are identified and controlled with small vascular clamps and divided. The veins are ligated on the mesenteric vein side and the portal vein side with 5-0 polypropylene sutures with care taken to avoid narrowing of the superior mesenteric vein or the portal vein.

Step 19. The uncinate process is retracted to the right, the portal vein retracted to the left, and the superior mesenteric artery identified by visualization and palpation. Small branches of the superior mesenteric artery into the uncinate process are identified and sequentially divided between fine clamps. The uncinate process is completely dissected off the lateral surface of the superior mesenteric artery, and the branches of the superior mesenteric artery are controlled with 4-0 and 3-0 polypropylene suture ligatures.

20

Step 20. The specimen is removed and sent to pathology for frozen sections on the bile duct margin, the uncinate process margin, and the pancreatic neck margin. Titanium clips are placed along the retroperitoneal margin on the lateral edge of the superior mesenteric artery for postoperative localization by radiation oncology. The cut end of the left pancreas is then retracted superiorly and the posterior branches from the superior mesenteric vein and the splenic vein are divided between clamps and oversewn with 5-0 polypropylene suture ligatures. Approximately 1 inch of the proximal body of the pancreas is mobilized for preparation of the anastomosis.

Step 21. The proximal end of the jejunum is then brought through an opening in the transverse mesocolon to the right of the middle colic vessels. In this retrocolic position, an end-to-side pancreatic jejunal anastomosis is done. The end of the pancreas is sewn to the antimesenteric border of the jejunum in two layers. An outer layer of interrupted 4-0 polypropylene sutures and a similar inner layer of interrupted 4-0 polypropylene sutures are placed from the posterior surface working anteriorly. When possible, separate sutures are placed in the corners of the pancreatic duct and then stitched to the opening in the small bowel in a full-thickness layer. This helps maintain patency of the pancreatic duct and minimizes anastomotic leaks.

Step 22. The next anastomosis performed is the hepaticojejunostomy. This is also performed in an end-to-side fashion between the end of the common hepatic duct and the antimesenteric border of the jejunum. This anastomosis is placed approximately 10-15 cm distal to the pancreatic anastomosis and is performed in a single layer with interrupted 4-0 absorbable monofilament sutures. The opening in the transverse mesocolon is reapproximated with interrupted 3-0 absorbable sutures.

Step 23. A side-to-side gastrojejunal anastomosis is then placed approximately 30-35 cm distal to the biliary anastomosis. This may be placed in a retrocolic or antecolic position, but the antecolic position is preferred as it may minimize postoperative radiation injury to the anastomosis or subsequent blockage from local recurrent cancer.

Step 24. Two 10 mm flat closed suction drains are placed in the right upper quadrant. An inferior drain is placed posterior to the biliary anastomosis, and a superior drain is placed between the pancreatic anastomosis and the incision.

Step 25. A feeding jejunostomy may be placed approximately 30 cm distal to the gastrojejunal anastomosis. The abdomen is irrigated, hemostasis is assured, the abdominal wall is closed in layers, and the skin may be closed with staples for expediency.

Postop

Nasogastric decompression is maintained until bowel function resumes. Fluid amylase levels are checked on the drain output on postoperative day 6 and if low, drains are pulled on or about postoperative day 7. Glucose levels are monitored for postoperative insulin requirements. Pancreatic enzyme replacement may be required if there is evidence of pancreatic exocrine function insufficiency. Erythromycin 500 mg intravenously every 8 hours may be started on postoperative day 3 as a motility agonist. Deep venous thrombosis prophylaxis with subcutaneous heparin and sequential compression devices is essential. Antibiotics are discontinued after the 24-hour perioperative period.

Complications

Early complications include postoperative hemorrhage, which may be from the gastric anastomosis, the cut edge of the pancreas, or from branches of the superior mesenteric artery or superior mesenteric vein. Subsequent complications include leakage from the pancreaticojejunal anastomosis or the hepaticojejunal anastomosis, delayed gastric emptying, and postoperative marginal ulcer formation.

Follow-Up

Patients should be seen frequently in the weeks following surgery to assess healing and return of functional status. Insulin and/or pancreatic enzyme replacement may be required. Adjuvant chemoradiation therapy is ordinarily given unless it was given preoperatively. Since the likelihood of recurrent disease is high, patients should be seen regularly and repeat imaging performed if dictated by a deterioration in clinical status.

20

Chapter 21

Distal Pancreatectomy and Splenectomy

Woody Denham

Indications

Distal pancreatectomy is indicated for neoplasms of the body or tail of the pancreas, for the treatment of symptomatic chronic pancreatitis limited to the pancreatic body/tail, and for the treatment of chronic pancreatic fistula/pseudocyst arising from the distal pancreas.

Preop

An epidural catheter for postoperative pain control should be placed preoperatively unless contraindicated. Deep venous thrombosis prophylaxis with sequential compression devices and/or subcutaneous heparin (depending on patient risk factors) should be instituted prior to the initiation of general anesthesia. A prophylactic perioperative antibiotic is administered intravenously 30 minutes prior to incision.

Procedure

Step 1. A bilateral subcostal incision is preferred. A midline incision is also acceptable.

Step 2. The lesser sac is entered by elevating the omentum off the transverse colon, and the body and tail of the pancreas is exposed.

Step 3. The peritoneum along the inferior border of the pancreas is incised. The body and tail of the pancreas can be elevated by gentle dissection behind the gland.

Step 4. The short gastric vessels between the spleen and stomach are ligated and divided starting at the midportion of the greater curve of the stomach and moving upward. The highest short gastric vessels can be ligated once the spleen has been mobilized (Step 5) if exposure is difficult.

Step 5. The lateral peritoneal attachments of the spleen are divided and the spleen mobilized anteriorly. Splenic attachments to the colon and diaphragm must be divided. The attachments to the colon may require ligation and division due to their vascular nature. Any residual short gastric vessels spanning from the spleen to the stomach are divided so that the stomach can be completely retracted to the patient's right.

Step 6. The pancreas and spleen are mobilized upwards and towards the patient's right by blunt dissection between the kidney and the tail of the pancreas. After some mobilization, the hand meets the retropancreatic space already created in Step 3.

Step 7. The inferior mesenteric vein, if encountered, is ligated and divided as it joins the splenic vein. Disecction of the peripancreatic soft tissues continues until the confluence of the splenic vein and superior mesenteric vein is visualized.

Northwestern Handbook of Surgical Procedures, edited by Richard H. Bell, Jr. and Dixon B. Kaufman. ©2005 Landes Bioscience.

Step 8. Along the upper border of the pancreas, the splenic artery is identified near its origin from the celiac axis before it enters the pancreatic substance. It is doubly ligated proximally with 2-0 suture, ligated once distally, and divided.

Step 9. The splenic vein is clamped with vascular clamps and divided. The portal end of the vein is oversewn with continuous 5-0 polypropylene suture. The splenic end can be ligated with a 2-0 suture ligature.

Step 10. The pancreatic parenchyma is divided using electrocautery. In patients with a neoplasm of the pancreas, a margin of at least 1 cm must be present.

Step 11. A 3-0 monofilament nonabsorbable suture is used to ligate the main pancreatic duct. Interrupted horizontal mattress sutures are placed through the edge of the pancreatic parenchyma to close the cut edge.

Step 12. A soft, closed-suction drain is placed adjacent the cut edge of the pancreas and brought out through the left lateral abdominal wall.

Postop

A nasogastric tube is usually left for 24 hours, then diet is begun gradually. The closed-suction drain is removed after oral intake is resumed if there is low output and the amylase level in the drain fluid is not elevated.

Complications

The most significant complications of distal pancreatectomy and splenectomy are pancreatic fistula from the cut edge of the gland and/or left subphrenic abscess.

Follow-Up

In the early postoperative weeks, the platelet count should be monitored. If it rises above 1,000,000/mm^3, aspirin therapy should be initiated. Patients should also be followed long-term for the development of diabetes or steatorrhea. This is particularly true for patients who undergo distal pancreatectomy for chronic pancreatitis. In this group, up to 33% develop postoperative diabetes and 20% may develop steatorrhea.

21

Chapter 22

Pancreatic Cystogastrostomy

Richard H. Bell, Jr.

Indications

Pancreatic cystogastrostomy is indicated for the treatment of large, persistent, and/or symptomatic pseudocysts of the pancreas. The pseudocyst should indent the posterior wall of the stomach by CT or other imaging.

Preop

Intraoperative ultrasound should be available if necessary. An epidural catheter for postoperative pain control is placed immediately prior to surgery. An intravenous antibiotic for surgical site infection prophylaxis (cefazolin 1.0 g IV) is given 30 minutes prior to incision. Pneumatic compression boots and/or subcutaneous heparin is used for deep venous thrombosis prophylaxis according to patient risk. A nasogastric (NG) tube and urinary catheter are placed after induction of anesthesia.

Procedure

Step 1. An upper midline abdominal incision is made. The abdomen is explored to the extent possible.

Step 2. The pseudocyst is palpated through the stomach. Using its location as a guide, a transverse incision about 5 cm long is made through the anterior gastric wall with electrocautery.

Step 3. Using narrow Deaver retractors or sutures for retraction of the edges of the stomach incision, the point where the pseudocyst bulges maximally into the posterior gastric wall is established by inspection and palpation.

Step 4. A 20-gauge needle attached to a 10 cc syringe is passed through the posterior gastric wall, aspirating continuously. Pseudocyst fluid should be encountered within 1-2 cm. If no fluid is encountered, aspiration in a different site or intraoperative ultrasound should be considered. The posterior wall of the stomach must not be opened unless fluid is encountered by aspiration within a short distance of the gastric wall.

Step 5. If fluid is encountered, an incision is made through the posterior wall of the stomach with the electrocautery. Initially, a 2-3 cm incision is made to enter the pseudocyst. The incision may be enlarged depending on the size of the pseudocyst. The final incision should be at least 3-4 cm long. It is inadvisable to palpate the inside of the pseudocyst vigorously, as bleeding may develop which is difficult to control.

Step 6. Using the electrocautery, a thin ellipse of tissue is removed from the edge of the incision into the pseudocyst which encompasses some of the wall of the pseudocyst. The tissue must be sent to pathology for frozen section to rule out a

Northwestern Handbook of Surgical Procedures, edited by Richard H. Bell, Jr. and Dixon B. Kaufman. ©2005 Landes Bioscience.

cystic neoplasm of the pancreas, which will differ from a pseudocyst by possessing an epithelial lining.

Step 7. The edge of the cystogastrostomy incision should be checked for bleeding, which should be controlled with cautery or suture ligatures.

Step 8. Starting a separate suture at each corner, the anterior wall of the stomach is closed with an inner layer of continuous absorbable 3-0 suture, incorporating all layers of the stomach wall. The closure is completed with an outer layer of interrupted seromuscular 3-0 silk sutures.

Step 9. The midline fascia and skin are closed with suture material of choice.

Postop

The NG tube can be removed on postoperative day 1 and diet begun as tolerated.

Complications

The most significant procedure-specific complication is gastrointestinal bleeding from the cystogastrostomy. This occurs in approximately 10% of patients. Bleeding should be evaluated with upper endoscopy and also by angiography to rule out a pseudoaneurysm associated with the pseudocyst, which can be embolized.

Follow-Up

After recovery from surgery, patients should be followed at intervals to determine if any symptoms recur. Recurrence of symptoms should prompt imaging studies of the abdomen to rule out pseudocyst recurrence, which occurs in about 10% of cases, primarily in the first year after surgery.

22

Chapter 23

Longitudinal Pancreaticojejunostomy: Puestow Procedure

Richard H. Bell, Jr.

Indications

Longitudinal pancreaticojejunostomy is indicated in patients with chronic pancreatitis with refractory pain, who have failed symptomatic therapy, whose entire pancreatic duct is dilated, and whose gland is generally atrophic.

Preop

Intraoperative ultrasound should be requested. The patient should be NPO for 8 hours preoperatively. Bowel prep is not necessary. An intravenous antibiotic for surgical site infection prophylaxis (cefazolin 1 g IV) is given 30 minutes before incision. A nasogastric (NG) tube and urinary catheter are placed after general anesthesia is induced. Sequential calf compression boots and/or low-dose subcutaneous heparin are administered for deep vein thrombosis (DVT) prophylaxis depending on patient risk factors.

Procedure

Step 1. A bilateral subcostal incision is made extending to the lateral edge of the rectus abdominis muscle on each side. A general exploration of the abdomen is performed. The NG tube position is confirmed.

Step 2. An upper abdominal retractor system is placed.

Step 3. The hepatic flexure of the colon is mobilized by incising the peritoneum above the flexure and the base of the transverse mesocolon is swept inferiorly to expose the duodenum and head of the pancreas.

Step 4. After incising the peritoneum lateral to it, the duodenum is mobilized with a Kocher maneuver extending from the right edge of the hepatoduodenal ligament to the superior mesenteric vessels.

Step 5. The greater omentum is dissected upward and away from the transverse colon and the lesser sac thereby entered. The stomach is mobilized in a cephalad direction by blunt and sharp dissection and the stomach and omentum retracted upward to allow identification of the body and tail of the pancreas. The entire anterior surface of the body and tail of the gland should be cleaned and exposed.

Step 6. On the upper edge of the transverse mesocolon, the middle colic vein is followed down to its entrance into the superior mesenteric vein (SMV). The right gastroepiploic vein, which usually joins the middle colic vein to form the gastrocolic trunk before they enter the SMV, is identified. The right gastroepiploic vein is then ligated and divided as close to its entrance into the gastrocolic trunk or SMV as

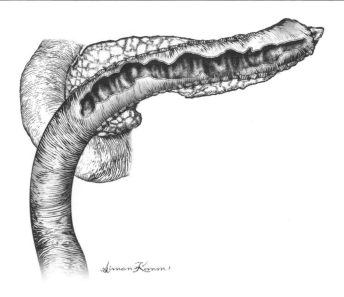

Figure 23.1. Longitudinal pancreaticojejunostomy.

possible, taking care not to injure the middle colic vein. Because the right gastroepiploic vein crosses the neck of the pancreas from top to bottom to enter the SMV, it must be divided to provide adequate exposure of the junction between the head and body of the pancreas.

Step 7. With the entire anterior surface of the pancreas well exposed from the head to the tail, the ultrasound probe is applied to an area of the gland that is easily accessible and the pancreatic duct identified. A 20-gauge needle is passed into the duct, aspirating with a syringe to confirm the return of pancreatic juice. The needle is left in place once juice is returned and the electrocautery used to cut down on the duct and enter it.

Step 8. The duct is opened as far as possible in both directions by passing a right-angle clamp into it and progressively dividing overlying parenchyma with electrocautery. To the patient's right, the duct should be opened to within 1-2 cm of the duodenum. To the left, the duct should be opened along the body and tail of the gland to within 3-4 cm of the tail of the gland. The gland should not be opened so far to the left that subsequent anastomosis will be overly difficult due to poor visibility.

Step 9. Once the pancreatic dissection is complete, an appropriate point in the proximal jejunum is identified to create a Roux-en-Y limb and the jejunum divided with a linear stapler. The mesentery of the jejunum is divided perpendicular to the axis of the bowel to allow mobilization of the limb. The distal end of the jejunum is oversewn with interrupted sutures of 3-0 silk. A small defect is created in the mesentery of the transverse colon and the distal end of the divided jejunum passed through the defect, bringing the end of the jejunum to the patient's left. The Roux limb should lie easily over the opened pancreas without tension.

23

Step 10. The Roux limb is sewn to the inferior border of the pancreas with interrupted sutures of 3-0 silk which are placed into the seromuscular layers of the bowel and into the pancreatic parenchyma. The small bowel is then opened with cautery, making the opening parallel to the pancreatic duct but keeping the opening in the bowel slightly shorter than the opening in the pancreas.

Step 11. Starting in the middle of the back row of the anastomosis, a double-armed suture of 3-0 nonabsorbable material is placed and then a circumferential anastomosis of all bowel layers to the cut edge of the pancreatic defect is performed. The anastomosis is completed with an anterior row of interrupted 3-0 silk sutures.

Step 12. An end-to-side jejunojejunostomy is performed 40 cm below the pancreatic anastomosis to complete the Roux limb and the leaves of the mesentery sutured to prevent internal hernia. The edge of the defect in the transverse mesocolon is also sutured to the serosa of the Roux limb.

Step 13. Closed-suction drains are placed both above and below the anastomosis and brought out through the abdominal wall..

Step 14. The abdominal wall is closed in layers with suture material of choice.

Postop

The NG tube can be removed on postoperative day 1. Diet is begun as tolerated. The drains are removed after the patient takes a diet if no amylase-rich drainage is present.

Complications

The most important procedure-specific complication is leakage from the pancreaticojejunostomy, which occurs in under 5% of patients. If a leak occurs, the patient should be converted to total parenteral nutrition, which may be continued at home if the leak is controlled. If the leak is associated with sepsis or bleeding from drains, reexploration may be required.

Follow-Up

After recovery, patients should be followed indefinitely at regular intervals (6-12 months) to determine if pain is relieved and to determine if additional complications of chronic pancreatitis have developed, such as diabetes, exocrine insufficiency, or left-sided portal hypertension.

23

Chapter 24

Duodenum-Preserving Subtotal Pancreatic Head Resection: Frey Procedure

Richard H. Bell, Jr.

Indications

Duodenum-preserving subtotal resection of the pancreatic head is indicated in patients with chronic pancreatitis with refractory pain not responsive to nonsurgical therapy, whose pancreatic head is enlarged (greater than 7 cm in diameter), and whose distal pancreatic duct is dilated. The procedure is also applicable if patients have associated compression of the common bile duct or duodenum.

Preop

Intraoperative ultrasound should be requested. Bowel prep is not necessary. An intravenous antibiotic for prophylaxis of surgical site infection (cefazolin, 1 g IV) is given 30 minutes before incision. A nasogastric (NG) tube and urinary catheter are placed after general anesthesia is induced. Pneumatic calf compression boots and/or low-dose subcutaneous heparin are used for deep vein thrombosis (DVT) prophylaxis according to patient risk profile.

Procedure

Step 1. A bilateral subcostal incision is made extending through the rectus abdominis muscle on each side. The abdomen is explored. NG tube position is confirmed.

Step 2. An upper abdominal retractor system is placed.

Step 3. The hepatic flexure of the colon is mobilized and the base of the transverse mesocolon swept inferiorly to expose the duodenum and head of the pancreas.

Step 4. By incising the peritoneum lateral to the duodenum, a Kocher maneuver is performed, mobilizing the duodenum from the lateral edge of the hepatoduodenal ligament to the superior mesenteric vessels.

Step 5. If possible, the anterior surface of the superior mesenteric vein (SMV) is identified at this point by incising the peritoneum and fatty tissue overlying the vein. The vein is followed to the point where it passes behind the neck of the pancreas. It is not necessary to dissect behind the neck of the pancreas. When there is significant peripancreatic inflammation present, dissection of the vein may not be possible at this time and the vein can be identified later.

Step 6. The greater omentum is dissected upward and away from the transverse colon and the lesser sac entered. The stomach is mobilized in a cephalad direction by blunt and sharp dissection and the stomach and omentum retracted upward. The body and tail of the pancreas are identified and the entire anterior surface exposed.

Northwestern Handbook of Surgical Procedures, edited by Richard H. Bell, Jr. and Dixon B. Kaufman. ©2005 Landes Bioscience.

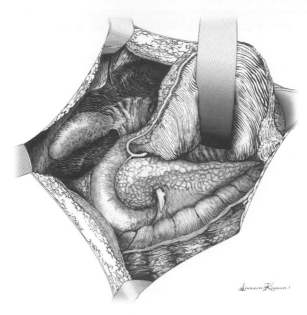

Figure 24.1. Duodenum-preserving subtotal pancreatic head resection. Pancreas exposure.

Step 7. On the upper surface of the transverse mesocolon, the middle colic vein is identified and followed down to its entrance into the SMV. If not done previously, the anterior surface of the SMV is cleaned by incising the peritoneum and fatty tissue overlying the vein. The vein is followed to the point where it passes behind the neck of the pancreas. The right gastroepiploic vein is identified, which usually joins the middle colic vein to form the gastrocolic trunk before the two veins enter the SMV. The right gastroepiploic vein is ligated as close as possible to its entrance into the gastrocolic trunk or SMV, taking care not to injure the middle colic vein. Because the right gastroepiploic vein crosses the neck of the pancreas from top to bottom to enter the SMV, it must be divided to provide adequate exposure of the junction between the head and body of the pancreas.

Step 8. After assuring that the entire anterior surface of the pancreas is well exposed from the head to the tail, the ultrasound probe is applied to an area of the gland that is easily accessible and the pancreatic duct identified. A 20-gauge needle is passed into the duct, aspirating into a syringe to confirm the return of pancreatic juice. The needle is left in place once juice is returned and the electrocautery used to cut down on the duct and enter it.

Step 9. The pancreatic duct is opened as far as possible in both directions by passing a clamp into the duct and progressively dividing the overlying parenchyma with electrocautery. To the patient's right, the duct should be opened to within 1-2 cm of the duodenum. To the left, the duct should be opened along the body and tail of the gland to within 3-4 cm of the tail of the gland. The duct should not be opened so far to the left that subsequent anastomosis will be overly difficult due to poor visibility.

24

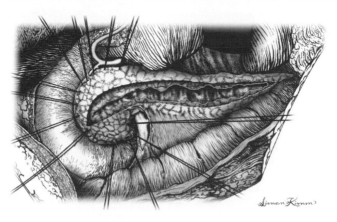

Figure 24.2. Duodenum-preserving subtotal pancreatic head resection. Pancreatic duct is open.

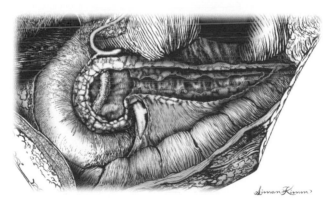

24

Figure 24.3. Duodenum-preserving subtotal pancreatic head resection. Partial resection of head of the pancreas.

Step 10. At this point, small portions of the head of the gland are resected by working from the opened duct outward. Pieces of the head are removed with a scalpel, obtaining hemostasis as needed with cautery or ligatures. The dissection can be quite bloody. The left hand is placed behind the head of the gland to guide the dissection. The objective is to leave a rim of the pancreatic head approximately 1 cm in thickness in all directions. If the bile duct is compressed, it is necessary to identify the duct within the head of the gland and decompress it from the point where it enters the gland to the point where it enters the ampulla. This dissection is easier if a metal probe (a Bakes dilator works well) is placed into the common duct through the cystic duct (necessitating cholecystectomy in patients who have not already had their gallbladder removed).

Step 11. Once the pancreatic dissection is complete, an appropriate point in the proximal jejunum is identified to create a Roux-en-Y limb and the jejunum divided with a linear stapler. The mesentery of the jejunum is divided perpendicular to the axis of the bowel to allow mobilization of the limb. A small defect is created in the mesentery of the transverse colon and the distal end of the divided jejunum passed through the defect, bringing the end of the jejunum to the patient's left. The Roux limb should lie easily over the opened pancreas without tension.

Step 12. The Roux limb is attached to the inferior border of the pancreas with interrupted sutures of 3-0 silk which are placed into the seromuscular layers of the bowel and into the pancreatic parenchyma. The small bowel is then opened, making the jejunotomy slightly smaller than the pancreatic defect.

Step 13. Starting in the middle of the back row of the anastomosis, a double-armed suture of continuous 3-0 nonabsorbable suture is placed, and then a continuous circumferential anastomosis of all bowel layers to the cut edge of the pancreatic defect is performed. The anastomosis is completed with a superior row of interrupted 3-0 silk sutures.

Step 14. A jejunojejunostomy is performed 40 cm below the pancreatic anastomosis to complete the Roux limb. The leaves of the mesentery are approximated to prevent internal hernia. The edge of the defect in the transverse mesocolon is likewise sewn to the serosa of the Roux limb.

Step 15. Closed-suction drains are placed both above and below the anastomosis.

Step 16. The abdominal wall is closed in layers with suture material of choice.

Postop

The NG tube can be removed on postoperative day 1 and diet instituted as tolerated. The drains are removed after the patient takes a diet if there is no amylase-rich drainage.

Complications

The most significant procedure-specific complication is leakage from the pancreaticojejunostomy, which occurs in under 5% of patients. If a leak occurs, the patient should be converted to total parenteral nutrition, which may be continued at home if the leak is controlled. If the leak is associated with sepsis or bleeding from drains, reexploration may be required.

Follow-Up

After recovery, patients should be followed indefinitely at regular intervals (6-12 months) to determine if pain is relieved and to determine if additional complications of chronic pancreatitis have developed, such as diabetes, exocrine insufficiency, or left-sided portal hypertension.

24

Chapter 25

Splenectomy: Open

Malcolm M. Bilimoria

Indications

Indications for splenectomy include blood dyscrasias refractory to medical management (i.e., autoimmune anemias, hereditary spherocytosis, immune thrombocytopenic purpura, and thrombotic thrombocytopenic purpura), primary splenic diseases (i.e., abscess or neoplasms), and trauma.

Preop

In patients undergoing elective splenectomy, immunization should be given at least 2 weeks preoperatively against pneumococcus, *H. influenzae*, and *N. meningitides*, all of which can cause overwhelming postsplenectomy sepsis. On the day of surgery, a preoperative prophylactic antibiotic is given. A nasogastric tube is placed. Deep vein thrombosis (DVT) prophylaxis should be provided by the use of sequential compression devices and/or subcutaneous heparin as appropriate for the patient's risk factors.

Procedure

Step 1. Either a midline or a left subcostal incision can be used.

Step 2. The lesser sac is entered through the left part of the gastrocolic omentum.

Step 3. Moving upward along the greater curvature of the stomach, the short gastric veins are divided with clamps and ties or divided with the aid of the harmonic scalpel. If the highest short gastric vessels are difficult to visualize, they can be ligated and divided after the spleen is mobilized.

Step 4. The splenocolic ligament between the spleen and the splenic flexure of the colon is divided, reflecting the colon downward.

Step 5. Retracting the spleen to the patient's right, the peritoneum lateral to the spleen is opened and the splenorenal and splenophrenic ligaments are divided. The operator's hand is passed behind the spleen and tail of the pancreas, mobilizing the spleen anteriorly and medially by blunt dissection until the spleen is brought up well anteriorly into the operative field.

Step 6. Any remaining short gastric vessels between the spleen and stomach are divided, retracting the stomach to the patient's right.

Step 7. The splenic artery is isolated just distal to the tail of the pancreas and ligated. Next the splenic vein is separately ligated and the spleen is removed.

Step 8. The retroperitoneum up into the left subphrenic space is thoroughly inspected for hemostasis.

Northwestern Handbook of Surgical Procedures, edited by Richard H. Bell, Jr. and Dixon B. Kaufman. ©2005 Landes Bioscience.

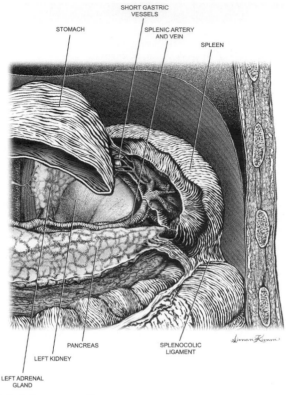

SHORT GASTRIC
VESSELS

STOMACH

SPLENIC ARTERY
AND VEIN

SPLEEN

PANCREAS

SPLENOCOLIC
LIGAMENT

LEFT KIDNEY

LEFT ADRENAL
GLAND

Figure 25.1. Splenectomy. Open.

Postop

A nasogastric tube is left in place overnight and can usually be removed on the first postoperative day. Diet is then advanced as tolerated. Platelet counts will peak at 5-7 days, and the use of aspirin for platelet counts greater than 600,000 is advisable.

Complications

A pancreatic leak may occur from damage to the tail of the pancreas and result in a subphrenic fluid collection. Necrosis of a portion of the greater curve of the stomach and leakage may occur if a ligature on one of the short gastric vessels incorporated part of the stomach wall. Arterial thrombosis and stroke have been reported secondary to the postoperative thrombocytosis.

Follow-Up

If the patient was operated for a hematologic indication, the response to therapy must be followed. In immune thrombocytopenic purpura, for example, there is a relapse rate of 15-20%. Postsplenectomy patients should generally be treated with prophylactic oral antibiotics when undergoing dental work or similar procedures that entail a risk of bacteremia.

Chapter 26

Splenectomy: Laparoscopic

Malcolm M. Bilimoria and Mark Toyama

Indications

Indications for splenectomy include blood dyscrasias refractory to medical management (i.e., autoimmune anemias, hereditary spherocytosis, immune thrombocytopenic purpura, and thrombotic thrombocytopenic purpura), and primary splenic diseases such as cysts or abscess.

Preop

The patient should receive immunization against pneumococcus, meningococcus, and *H. influenzae* 1-2 weeks preoperatively. Routine preoperative antibiotics are administered 30 minutes prior to incision. A nasogastric (NG) tube is placed to decompress the stomach during the procedure. Sequential compression stockings or subcutaneous heparin should be administered for patients at risk for deep vein thrombosis/pulmonary embolism.

Procedure

Step 1. The patient is placed in a 45° right lateral decubitus position.

Step 2. Obtain peritoneal access at or above the umbilicus, depending on the patient's body habitus, and insufflate in the standard fashion to a pressure of 15 mm Hg. Inspect the abdomen with a 30 or 45° laparoscope.

Step 3. Three or four additional ports are placed along the left costal margin. A combination of 5, 10, and 12 mm ports may be used depending on instrument preference. One of the lateral ports is usually a 12 mm port to accomodate the laparoscopic stapling device. An additional port is sometimes needed to retract the left lobe of the liver.

Step 4. Mobilize the splenic flexure of the colon with the harmonic shears by retracting it down and to the right and dividing the tissues between the colon and spleen.

Step 5. Continue with the posterior dissection of the spleen by dividing the splenophrenic and splenorenal ligaments.

Step 6. Enter the lesser sac by dividing the greater omentum between the stomach and spleen.

Step 7. Moving towards the top of the spleen, sequentially divide the short gastric vessels with the harmonic shears. Careful retraction of the gastric fundus and upper pole of the spleen will assist in dividing the highest short gastric vessels.

Step 8. Identify the hilum of the spleen and carefully dissect free the splenic artery and vein.

Northwestern Handbook of Surgical Procedures, edited by Richard H. Bell, Jr. and Dixon B. Kaufman. ©2005 Landes Bioscience.

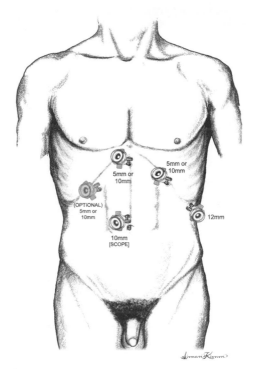

Figure 26.1. Splenectomy. Laparoscopic.

Step 9. If possible, divide the splenic artery first by passing the laparoscopic stapling device with a vascular staple load through a lateral port and positioning it perpendicular to the vessel.

Step 10. Divide the splenic vein with the stapling device in a similar fashion.

Step 11. Multiple firings of the stapling device may be necessary to divide all of the vascular attachments to the spleen.

Step 12. Divide the remaining posterior attachments of the spleen with the harmonic shears.

Step 13. Place the spleen into a large specimen bag and remove it by either morcellization through a large port site or by extending one port site incision and removing the spleen intact.

Step 14. Carefully inspect the splenic hilar area and omentum for any evidence of an accessory spleen(s).

Step 15. Irrigate the operative field and suction. Drain placement is not required unless there is concern for a pancreatic injury.

Step 16. Close the fascia at all port sites 10 mm in size or larger. Close the skin at all port sites.

26

Postop

The NG tube is left in place overnight and can usually be removed the first postoperative day. If the procedure was done for thrombocytopenia, coagulation status and platelet counts should be rechecked postoperatively. Discharge is usually possible within 24-48 hours.

Complications

Immediate complications include postoperative hemorrhage, a pancreatic leak from damage to the tail of the pancreas, or necrosis of a portion of the greater curvature of the stomach. Postoperative subphrenic abscess may occur. An arteriovenous fistula may develop between the splenic artery and vein stumps. Postsplenectomy sepsis occurs in approximately 1% of patients.

Follow-Up

Patients are usually seen 2-3 weeks after their procedure in routine postoperative follow-up. If the patient was operated for a hematologic condition, long-term follow-up with their hematologist may be necessary for monitoring and any further medical therapy.

26

Chapter 27

Splenorrhaphy: Open

Michael A. West

Indications

Splenorrhaphy is indicated when the grade of splenic injury requires repair, but does not require splenectomy. Minor injuries (e.g., grade I capsular tear) should be treated with local measures or topical hemostasis and do not require splenorrhaphy. In general, grade II and III spleen injuries would have the greatest likelihood of benefiting from splenorrhaphy. If possible, it is always desirable to preserve spleen mass and function to minimize the risk of late infection or immune defects.

Preop

Prior to performing exploratory laparatomy the patient should have appropriate venous access and should (if possible) be well resuscitated. Blood for type and cross-match should be sent. It is advantageous to place a Foley catheter prior to abdominal exploration. A nasogastric (NG) suction catheter should be placed preoperatively. When performing exploratory laparotomy for trauma, the surgeon should be sure that there are two suction devices and carefully position, prep, and drape the patient such that the chest and/or mediastinum can be accessed intraoperatively. Antibiotic prophylaxis should be instituted prior to making the incision. Choice of agent should be based on the pathogens likely to be encountered. A second-generation cephalosporin or other agents that cover aerobic and anaerobic enteric pathogens are frequently used.

Procedure

Step 1. Exploratory laparotomy is performed through a midline incision.

Step 2. A large body wall retractor is placed in the left upper quadrant to permit examination of the spleen. Initially examine the spleen in situ, looking for evidence of deep lacerations, active bleeding, or injury to hilar structures. It is not necessary to mobilize the spleen in the absence of splenic injury. Mobilization of the spleen may exacerbate bleeding from minor injuries. If there is evidence of significant splenic injury and onging hemorrhage, the spleen should be completely mobilized.

Step 3. Splenic mobilization is performed by retracting the spleen anteriorly and medially using the surgeon's nondominant hand. This maneuver places the splenorenal ligament "on stretch." This ligament can then be incised using a scissors or cautery, beginning at the inferior aspect of the spleen and continuing in a cephalad direction. In some instances this maneuver must be performed blindly, by "feel." The spleen is mobilized medially using predominantly blunt dissection in the plane posterior to the pancreas.

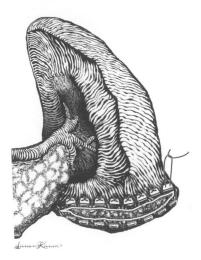

Figure 27.1. Splenorrhaphy. Horizontal mattress sutures with pledgets to control hemorrhage.

Step 4. The short gastric vessels in the splenogastric ligament should be divided. In many instances, the side of the vessel on the greater curve of the stomach should be suture-ligated to avoid dislodgment from postoperative gastric distension. The splenocolic ligament is likewise divided and the splenic flexure of the colon swept inferiorly. Occasionally this structure will contain significant vessels that require clamping. The mobilized spleen is rotated upward into the midline abdominal incision. This may be facilitated with placement of several laparotomy pads in the splenic fossa.

Step 5. Hemorrhage can be temporarily controlled (if needed) by application of digital pressure to the hilum of the mobilized spleen. Debridement should be performed very sparingly, if at all.

Step 6. Horizontal mattress sutures are employed to control hemorrhage and reapproximate isolated lacerations of the splenic parenchyma. Some type of pledget or "bolster" material is usually required to prevent tearing the splenic parenchyma with exacerbation of bleeding. Teflon felt, polyglycolic acid mesh, and omentum have been employed for this purpose. Absorbable sutures (2-0) on a large taper or blunt needle are carefully placed in an interlocking horizontal mattress fashion. In most instances, the sutures are best placed so that the knots are on the diaphragmatic surface.

Step 7. In instances where multiple or deeper lacerations are found the spleen can sometimes be salvaged by performing a wrap using polyglycolic acid mesh. The splenic wrap is performed using a large sheet of absorbable mesh material. A "keyhole" slit to accommodate the hilar vessels is cut on one side of the mesh. The mesh is passed behind the spleen and the mesh folded to envelop the organ. The free edges of the mesh are approximated with a running 2-0 absorbable suture. It is desirable to fashion the mesh and secure the closure to achieve a slight degree of compression. However, care must be taken to avoid vascular compromise or splenic infarction.

27

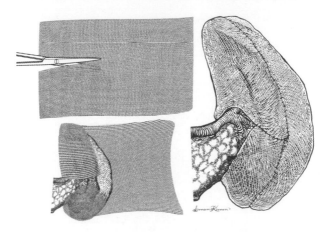

Figure 27.2. Splenorrhaphy. A splenic wrap with polyglycolic acid mesh.

Step 8. Anatomic resection may be employed if injury is localized to one pole of the spleen. The appropriate hilar vessel is ligated and horizontal mattress sutures are placed proximal to the line of vascular demarcation using the technique described in Step 6. After the sutures have been secured, the nonviable splenic parenchyma is resected distal to the suture line.

Step 9. The splenic bed is carefully inspected to ensure that hemostasis is complete prior to closure. Laparotomy sponges in the splenic fossa can be rolled anteriorly to assist the surgeon to visualize this area. Drains are not used.

Step 10. Abdominal fascial and skin closure are performed as described for exploratory laparotomy.

Additional Comments

Complete abdominal exploration should be performed prior to attempting splenorrhaphy. This is important to determine the presence of other injuries and to appropriately prioritize the operative treatment(s).

Postop

Careful postoperative hemodynamic management and fluid management are indicated following splenorrhaphy. Hypertension should be avoided. Routine postoperative CT scanning confers little benefit.

Complications

The most common complication of splenorrhaphy is rebleeding. Splenectomy should be performed if reoperation is required for ongoing bleeding. Infectious complications may occur in the left subphrenic area. Most fluid collections or abscesses can be drained percutaneously. Many surgeons feel that avoiding Teflon pledgets minimizes infectious complications after splenorrhaphy.

Follow-Up

No specific long-term follow-up is required. If splenic preservation was successful there is no need to administer polyvalent vaccines against encapsulated bacteria.

27

Chapter 28

Small Bowel Resection and Anastomosis (Enterectomy): Open

Joseph A. Caprini

Indications

Enterectomy with anastomosis is indicated in a variety of conditions, including congenital atresia or stenosis of the small bowel, blunt and penetrating injuries of the small bowel, benign and malignant small bowel tumors, bleeding Meckel's diverticulum, inflammatory bowel disease, intestinal fistula, intestinal gangrene, intussusception in adults, some cases of meconium ileus or intestinal duplication, mesenteric tumors that in the course of removal could produce vascular compromise to adjacent small bowel, and when small bowel is adherent to intraabdominal tumors arising in other organs.

Preop

Antibiotic coverage should be started 30 minutes before the operation if elective. Antibiotics may be continued postoperatively if there is significant spillage of intestinal contents during the resection, gangrene, or infection. Decompression of preoperative bowel obstruction with nasogastric intubation when possible is very important. General anesthesia is usually employed. Patients should receive deep venous thrombosis prophylaxis appropriate to their level of risk.

Procedure

Step 1. The patient lies supine on the operating table, and the abdomen can be explored through a variety of incisions, depending on the pathology.

Step 2. A thorough systematic exploration of the abdomen should be done in most cases. There are times when this is not feasible, but one must guard against partial examinations that can miss important pathology.

Step 3. The area of small bowel containing the pathology should be isolated along with the appropriate section of mesentery. Towels can be used to wall off the area from the general abdominal cavity. Noncrushing intestinal clamps can be applied at some distance from the proposed resection lines to minimize spillage of the intestinal contents.

Step 4. A small opening in the mesentery immediately adjacent to the bowel wall should be made at the proposed proximal and distal resection sites, being careful to adequately expose the serosa of the bowel wall at the proposed resection sites.

Step 5. The peritoneum overlying the mesenteric leaf should be scored with scissors to outline the proposed line of resection of the mesentery. This resection line should extend from the small aperture adjacent to the bowel wall in a U-shaped fashion toward the other proposed resection margin. During this step it

---

Northwestern Handbook of Surgical Procedures, edited by Richard H. Bell, Jr. and Dixon B. Kaufman. ©2005 Landes Bioscience.

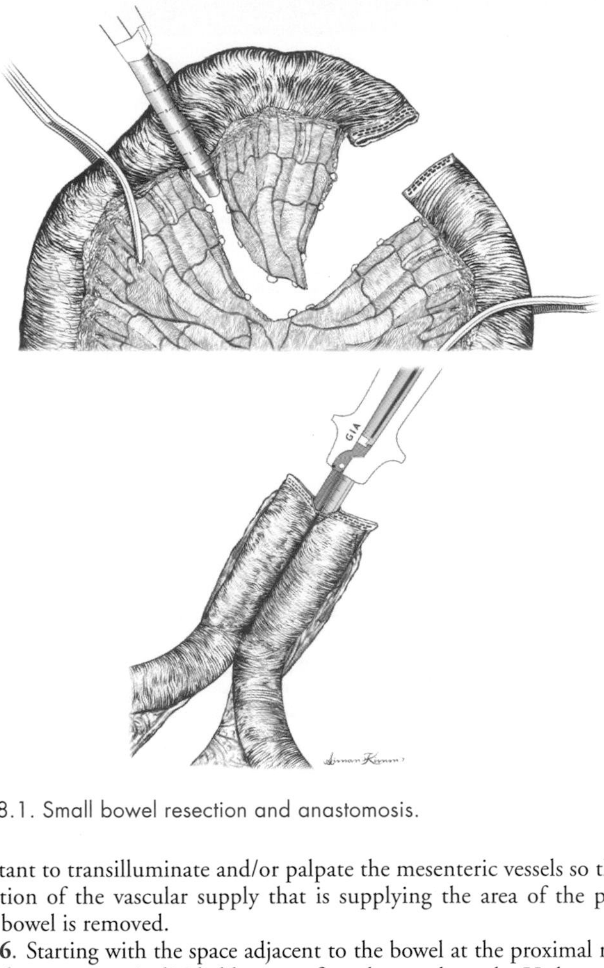

Figure 28.1. Small bowel resection and anastomosis.

is important to transilluminate and/or palpate the mesenteric vessels so that only that portion of the vascular supply that is supplying the area of the proposed resected bowel is removed.

Step 6. Starting with the space adjacent to the bowel at the proximal resection margin, the mesentery is divided between fine clamps along the U-shaped area to separate the mesentery of the resected bowel from the mesentery to be left behind. Mass ligatures should be avoided, and precise clamping and division of the individual vessels is preferred. Fine silk ligatures (4-0) are usually used depending on the size of the vessels. The extent of the mesenteric resection will depend on the amount of bowel to be removed. If a malignancy is present or suspected, an appropriate en bloc resection of mesenteric nodal tissue is required.

Step 7. The segment of the bowel to be removed is transected at each end with a linear stapler, which places two staggered rows of staples across the patient side and two more staggered rows on the specimen side of the bowel. The bowel is flattened and emptied, then the device is placed in a scissor-like fashion around the bowel wall and the staples fired. The built-in knife is then used to transect the bowel.

28

Step 8. The two ends of the bowel are placed adjacent to one another, and two seromuscular sutures are placed to approximate the bowel loops. Care must be taken not to twist the mesentery of the bowel. A small section of the antemesenteric border of the staple line closure of both bowel ends is excised. The bowel openings should be large enough to introduce one anvil of the linear stapler into each limb of bowel. One fork of the instrument is placed into each side of the bowel, and the bowel ends aligned evenly on the forks. The instrument is closed and the staples are fired. Then the knife is used to divide the tissue between the staple line which produces the stoma between the bowel loops. After the instrument is removed, it is very important to inspect the staple line for any bleeding.

Step 9. The common bowel opening is closed using a 55 mm double-row stapling instrument. One traction suture should be placed at each end of the bowel to hold the entire area in the staple instrument until the instrument is fired. Care must be taken to overlap the ends of the previous staple lines. One must avoid direct apposition of the anastomotic staple lines.

Step 10. The lumen is checked for patency and any leaks. The adequacy of the blood supply to the bowel at the suture lines is checked. The mesenteric opening is carefully closed with absorbable 4-0 suture, being careful not to injure the vessels to the bowel at the area of the anastomosis.

Additional Considerations

1. Gentle handling of the tissues is imperative with anastomosis of grossly healthy tissue only. Open anastomosis may be used and the risk of infection and later anastomotic leaks minimized by the use of appropriate antibiotics and limiting spillage of the intestinal contents at the time of anastomosis.

2. Adequate resection of the involved segment beyond the area of pathology is important in some diseases, including a 5-6 cm margin of tissue especially for malignant tumors.

3. It is good practice to cut the bowel obliquely, removing a greater portion of tissue on the antemesenteric side of the bowel. This is particularly important if the mesenteric circulation is borderline.

4. One must be very cautious when attempting an anastomosis in segments of bowel that are dilated. An increased incidence of leakage may occur depending upon the degree of dilatation.

Postop

The patients are not allowed to eat until they pass gas or otherwise show signs of return of bowel function. Nasogastric suction is used in the early postoperative period until bowel function is evident.

28

Complications

A variety of complications can occur which depend on the patient's disease and state of health. Among the most serious problems is a leak at the anastomosis, which typically would occur at about postoperative day 5-7. Postoperative ileus or bowel obstruction may occur and must be differentiated. Bleeding and wound infection also occur early.

Follow-Up

Long-term follow-up is dependent on the nature of the disease. Resection of short segments of small intestine has few if any long-term sequelae.

Chapter 29

Enterolysis for Small Bowel Obstruction: Open

Joseph A. Caprini

Indications

Enterolysis is ordinarily performed for small bowel obstruction in a patient with adhesions from previous abdominal surgery, past intraperitoneal infections, or congenital adhesive bands. Occasionally, although rarely, it is indicated for chronic or recurrent signs of intestinal obstruction including bloating, vomiting, and severe abdominal cramps if radiological studies support a diagnosis of mechanical obstruction.

Preop

Patients with intestinal obstruction may be volume depleted and require significant fluid resuscitation, although this should be done expeditiously. Decompression of preoperative bowel obstruction with nasogastric intubation as much as possible is very important. Antibiotic coverage should be started before the operation and may be continued postoperatively depending upon whether or not an inadvertent injury to the bowel occurred during division of adhesions. Patients should receive deep venous thrombosis prophylaxis appropriate to their level of risk.

Procedure

Step 1. The patient lies supine on the operating table. The abdomen may be explored through the previous incision, but it is often wise to slightly extend the incision so that one can enter the peritoneal cavity in normal tissue. In these patients, adhesions between the bowel and the abdominal wall are frequently present, and bowel injury may occur when the peritoneum is opened.

Step 2. A thorough, systematic exploration of the abdomen should be done as a routine. There are times when this is not feasible, but one must guard against partial examinations that can miss important pathology. The surgeon should look for the transition point between dilated and decompressed small bowel which marks the point of obstruction.

Step 3. Traction and countertraction in a gentle fashion are used to stretch out adhesions and allow sharp dissection. Gentle handling of the tissues is imperative, along with avoiding excessive tension when dividing adhesions. Adhesions should be divided in avascular planes. The use of sharp dissection is generally preferred for longstanding adhesions. Blunt tissue dissection can tear adjacent structures including the bowel. When extensive adhesions are present, one must be very cautious to work from areas of known anatomy into the unknown scarred areas. When working in the pelvis, one should be particularly careful to avoid the ureter and iliac veins. Bowel loops should be protected with saline-soaked gauze pads when they are temporarily lifted out of the abdominal cavity. Saline should be

Northwestern Handbook of Surgical Procedures, edited by Richard H. Bell, Jr. and Dixon B. Kaufman. ©2005 Landes Bioscience.

used at body temperature. Cold saline can cause hypothermia, and the peritoneum and bowel loops can be damaged by excessively warm saline.

Step 4. Examination of the entire small bowel is desirable, if possible, but is not absolutely necessary if the point of obstruction is obvious and dissection of the remaining bowel would be dangerous or excessively difficult. In patients with extensive adhesions, it is not necessary or desirable to take down all adhesions as long as it is clear that the adhesions causing the obstruction have been dealt with adequately. Areas of damage to the bowel serosa should be repaired with fine silk sutures. Once all of the adhesions have been divided as necessary to relieve the problem, the bowel is gently replaced in the abdominal cavity in as natural a position as possible.

Step 5. The abdominal cavity is irrigated with body-temperature saline, followed by closure of the abdominal wall with a large monofilament suture. An attempt should be made to cover the intestines with any available omentum to prevent bowel loops from coming into contact with the incision.

Postop

Nasogastric suction is used in the early postoperative period until bowel function is evident. The patients are allowed to initiate a diet once they pass flatus or otherwise show signs of return of bowel function.

Complications

Postoperative ileus and/or recurrent bowel obstruction may occur and must be differentiated. Wound dehiscence and wound infections are relatively common complications.

Follow-Up

The patient should be followed as an outpatient until wounds are healed. Patients with a small bowel obstruction are at significantly increased risk of a recurrent obstruction.

29

Chapter 30

Appendectomy: Open

Alexander P. Nagle

Indications

Appendectomy is indicated in patients for acute appendicitis. The patient's history and physical examination remain the most important aspects of diagnosing acute appendicitis. However, the common signs and symptoms such as fever, anorexia, and right lower quadrant pain are often absent; up to 33% of patients have an atypical presentation. Diagnostic accuracy varies by sex and age. The diagnosis of acute appendicitis may be difficult in women of childbearing age because symptoms of acute gynecologic conditions may manifest similarly. In selected clinical settings, a CT scan can improve the diagnostic accuracy up to 95%.

Preop

Correction of electrolytes, fluid resuscitation, and a Foley catheter are standard practice. Broad-spectrum antibiotics including adequate anaerobic coverage are administered. Both single- and multiple-agent therapy have been shown to be effective. Lower extremity sequential compression devices are used for deep venous thrombosis prophylaxis.

Procedure

Step 1. A right lower quadrant skin incision is made in the direction of the skin lines. The incision should be centered over McBurney's point (2/3 of the way along a line drawn from the umbilicus to the anterior superior iliac spine).

Step 2. The external oblique fascia is divided. The three muscle layers of the abdominal wall are split in the direction of their fibers and retracted. The peritoneum is opened transversely.

Step 3. As the peritoneal cavity is entered, any fluid should be noted, aspirated, and, if purulent, sent for a gram stain and culture. Any remaining fluid is then suctioned from the field.

Step 4. The cecum is identified, grasped, and exteriorized through the wound using a gauze sponge or blunt bowel graspers. The base of the appendix is identified at the confluence of the teniae coli. Often a finger can be used to sweep the appendix into the operative field. If the appendix is retrocecal, the cecum should be mobilized medially by incising the lateral peritoneal attachments of the cecum.

Step 5. Using a curved hemostat, a mesenteric window is created in an avascular area of the mesoappendix near the base of the appendix. The mesentery of the appendix is then divided between hemostats. It is important to correctly identify the appendiceal artery and assure that it is properly ligated. Any attachments at the appendiceal base are dissected in order to clearly identify the site of transection of the appendix.

Step 6. The appendix is divided approximately 5 mm from the cecal wall using a suture ligature. A clamp is placed distal to the suture ligature. Using a sharp scalpel, the appendix is transected between the suture ligature and the clamp. The mucosa of the appendiceal stump is lightly touched with electrocautery.

Step 7. In some cases of severe acute necrotizing appendicitis, the base of the appendix may not be suitable for transection and it may be necessary to perform a partial cecectomy.

Step 8. It is important to inspect the cut edge of the mesoappendix and appendiccal stump for integrity and hemostasis.

Step 9. The lower abdominal cavity is irrigated thoroughly with normal saline. No drains are required for cases of acute appendicitis.

Step 10. Closure of the fascial layers of the wound is carried out using running or interrupted absorbable suture. The skin and subcutaneous tissue are usually closed in acute appendicitis. In gangrenous and perforated appendicitis, the wound may be left open.

Special Considerations

The debate continues regarding the management of a normal-appearing appendix found at the time of surgery for presumed appendicitis. There are three possible situations: (1) normal-appearing appendix and no other pathology, (2) normal-appearing appendix and medically treated pathology (e.g., inflammatory bowel disease or pelvic inflammatory disease), and (3) normal-appearing appendix and surgically treated pathology (e.g., acute cholecystitis or perforated duodenal ulcer). The appendix should be removed in all but the last circumstance. This eliminates confusion should the patient return with similar symptoms.

Postop

Uncomplicated acute appendicitis requires minimal postoperative care. A significant decrease in pain and the return of bowel function can be anticipated by the day following operation. Diet is instituted as tolerated. Most patients can be discharged on postoperative day one. Oral opiates are used for pain control. For patients with a perforated appendix, antibiotics are generally continued for at least 5 days.

Complications

Acute infectious complications include abdominal abscess formation, cecal fistulas, and wound infections. The most important determinant of wound infection is the severity of contamination at the time of surgery. Wound infection rates of 1-4% occur with negative exploration. The risk increases to 5-12% with gangrenous appendicitis and 20-50% with perforated appendicitis. Antibiotic coverage for anaerobic organisms and a policy of open wound management after severe contamination reduce the risk of wound infection substantially. The mortality of acute appendicitis is contingent upon the state of the appendix at exploration and the age of the patient. Nonperforated appendicitis carries a 0.1% mortality rate and most deaths are due to intercurrent illness. Perforated appendicitis has a mortality rate of approximately 1% overall, usually secondary to infectious complications. In the elderly population, high rates of perforation and intercurrent illness contribute to a mortality rate in excess of 5%.

Follow-Up

The patient should return to see their surgeon 10-14 days after their operation. At that time, the pathology report should be reviewed; the presence of malignancy and carcinoid should be excluded. If no cancer is found within the specimen, no further follow-up is needed.

Chapter 31

Appendectomy: Laparoscopic

Mark Toyama

Indications

Laparoscopic appendectomy is indicated for the treatment of acute appendicitis or for interval appendectomy following nonoperative treatment of complicated appendicitis.

Preop

Standard intravenous antibiotic prophylaxis against wound infection is given preoperatively. The patient is placed in the supine position. For female patients, lithotomy position may be chosen, which allows manipulation of pelvic organs. A Foley catheter is inserted in the urinary bladder after the induction of anesthesia. Arms are tucked at the patient's side. The surgeon stands on the patient's left side.

Procedure

Step 1. The peritoneal cavity is accessed at the umbilicus (using either a Veress needle or a Hasson cannula), and a 5 mm or 10 mm trocar and laparoscope are placed.

Step 2. A 12 mm trocar is placed in the midline above the symphysis pubis. The incision can be hidden in the pubic hairline.

Step 3. A 2 mm or 5 mm trocar is placed in the right upper quadrant.

Step 4. An additional trocar can be placed as needed to improve exposure.

Step 5. The patient is then placed in Trendelenburg position and rotated towards the left to help expose the cecum.

Step 6. Instruments are placed and the appendix inspected.

Step 7. If the appendix is normal, a search for other sources of right lower quadrant pain should be undertaken, including adnexa in females, distal small bowel, gallbladder, etc.

Step 8. If the appendix is to be removed, it is elevated with an instrument and dissected free of any adhesions. It is often helpful to start at the appendiceal base.

Step 9. The appendix is elevated with a grasper or a pre-tied ligature to expose the mesoappendix.

Step 10. The mesoappendix is isolated with a dissector by making a window between it and the appendix.

Step 11. The mesoappendix is divided with clips, harmonic shears, or a laparoscopic stapling device.

Step 12. The appendix is divided at its base with a laparoscopic linear stapling device or with pre-tied ligatures.

Step 13. The appendix is placed in a specimen bag and removed through the largest port site.

Northwestern Handbook of Surgical Procedures, edited by Richard H. Bell, Jr. and Dixon B. Kaufman. ©2005 Landes Bioscience.

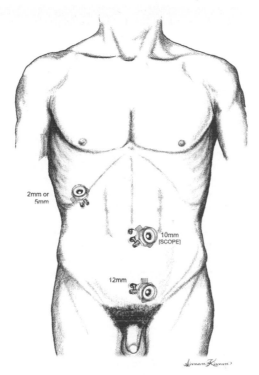

Figure 31.1. Laparoscopic appendectomy. Abdominal trocar placement.

Step 14. The abdomen is irrigated and suctioned. The appendiceal base and the mesoappendix should be inspected for hemostasis. The fascia of all port sites 10 mm in size or larger is closed.

Postop

Diet may be started when the patient is awake and alert if not nauseated. Discharge is often possible in less than 24 hours.

Complications

Postoperative complications include hemorrhage, periappendiceal abscess formation, wound infection, appendiceal stump leak with fistula formation. Incomplete appendectomy may become manifest much later with recurrent symptoms.

Follow-Up

Follow-up should occur in 1 to 2 weeks after discharge to inspect the incisions and review the pathology report.

31

Chapter 32

Ileostomy: Open Loop

Steven J. Stryker

Indications

The most common indication for loop ileostomy is to protect a low pelvic anastomosis such as a coloproctostomy, coloanal anastomosis, or ileoanal anastomosis. Other indications include proximal diversion for an unresectable, obstructing malignancy, for a radiation stricture, for pelvic sepsis from an anastomotic leak or trauma, or for severe anorectal Crohn's disease.

Preop

For elective resectional surgery that incorporates a complementary loop ileostomy, a standard mechanical bowel prep along with intraluminal and perioperative intravenous antibiotics should be employed. The patient should meet with an enterostomal therapist to begin ileostomy instruction and to have the stoma site selected and marked.

Procedure

Step 1. The patient is positioned on the operating table as dictated by the access required for any associated procedure. (i.e., supine vs. synchronous combined lithotomy). The abdomen is prepped and draped and the stoma site mark is scratched with a sterile needle to facilitate subsequent identification.

Step 2. If the operation requires a bowel resection or drainage of pelvic sepsis, this is carried out first.

Step 3. Just prior to closing the abdominal incision, the segment of ileum to be exteriorized is chosen. Typically, this is 10-15 cm proximal to the ileocecal junction. When the loop ileostomy is used in conjunction with proctocolectomy and ileoanal anastomosis, the surgeon should identify the most distal segment of ileum which has sufficient mesenteric laxity to comfortably reach the surface of the abdomen at the previously marked stoma site.

Step 4. An oval of skin and the underlying subcutaneous fat is excised at the stoma site. The anterior rectus fascia is incised in a cruciate fashion and the loop of ileum is drawn through the abdominal wall, passing through the rectus abdominis muscle. Care is taken to maintain the proper orientation of the loop and avoid twisting or angulation when exteriorizing the bowel.

Step 5. The loop is supported at the skin surface with a plastic rod slipped through a small opening in the mesentery adjacent to the bowel wall.

Step 6. The primary abdominal incision is closed completely. Then the loop ileostomy is matured by incising the distal limb transversely for about three quarters of its circumference and folding it back over the proximal limb. The full thickness of the opened edge of bowel wall is then sewn to the dermis circumferentially with absorbable suture.

Step 7. A stoma faceplate is cut to the corresponding size and placed over the newly fashioned stoma.

Postop

The patient is fed upon resumption of bowel activity. When the ileostomy begins to pass stool and flatus, the enterostomal therapist begins instruction in the care of the stoma. The peristomal skin is monitored for signs of irritation, which can be due to a poorly fitting stomal appliance. The skin margin adjacent to the distal limb opening is particularly prone to irritation with a loop ileostomy.

Complications

Early complications of loop ileostomy include bleeding from the mucocutaneous junction, peristomal wound infection, and stomal retraction. Later complications include prolapse, peristomal hernia, fistula, or internal hernia.

Follow-Up

If the loop ileostomy is formed to protect a distal anastomosis, it is best to wait about 10 weeks before closure. Prior to closing the stoma, a contrast radiograph of the distal anastomosis is obtained to assess the adequacy of healing. If the ileostomy was created because of intraabdominal or pelvic pathology, the stoma closure is deferred until the primary process has resolved.

32

Chapter 33

Hemicolectomy (Right): Open

David J. Winchester

Indications

Open right hemicolectomy is indicated for treatment of selected benign and malignant diseases of the colon and appendix.

Preop

The patient should ideally undergo a full mechanical bowel preparation and oral antibiotic bowel prep to minimize the risk of postoperative complications. In emergency settings, this may not be possible. In these circumstances, it is usually safe to proceed with resection and primary anastomosis due to the generous blood supply of the small bowel and relatively low bacterial count of the proximal colon. A nasogastric tube need not be placed unless the patient presents with an obstruction. Prophylactic antibiotics should be given 30 minutes prior to the incision. A Foley catheter should be inserted.

Procedure

Step 1. The patient is placed in the supine position. After induction of general anesthesia, a midline incision is made that extends above and below the umbilicus.

Step 2. The abdomen is explored. Under elective circumstances, when operating for colon cancer, perform an intraoperative ultrasound of the liver.

Step 3. Place a self-retaining retractor.

Step 4. Incise the peritoneal reflection of the right colon with electrocautery. Continue the mobilization distally to include the hepatic flexure. It may be necessary to ligate a few blood vessels at the hepatic flexure.

Step 5. Mobilizing the colon anteriorly, expose the retroperitoneum and identify the right ureter and duodenum. Avoid using electrocautery near these structures.

Step 6. Mobilize the ileum distally as it joins the cecum.

Step 7. Transect the ileum approximately 10 cm proximal to the ileocecal valve using an automated linear cutting stapler.

Step 8. Transect the colon just distal to the hepatic flexure using a second load on the stapling device.

Step 9. Score the peritoneum covering the mesocolon between the two points of transection. This may be done with electrocautery. This should be carried down to the origin of the right colic artery if the operation is being performed for cancer. For benign conditions, a lymphadenectomy is not necessary and may increase the chance of injuring a retroperitoneal structure such as the ureter.

Step 10. Clamp, ligate, and divide the mesenteric vessels with 2-0 silk suture. Other devices such as the Ligasure[R] or vascular staplers may be used but will increase the expense of the procedure.

Northwestern Handbook of Surgical Procedures, edited by Richard H. Bell, Jr. and Dixon B. Kaufman. ©2005 Landes Bioscience.

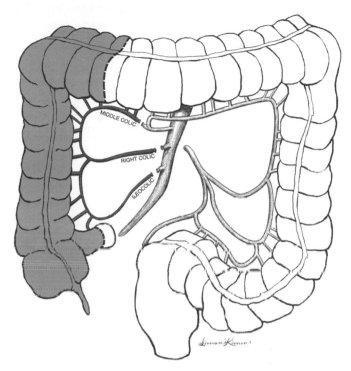

Figure 33.1. Right hemicolectomy.

Step 11. Remove the specimen from the field and provide orientation for the pathologist. Open the specimen to inspect the pathology and to be sure the suspected lesion has been removed with an adequate margin.

Step 12. Create a functional end-to-end anastomosis by firing a third load of the automated stapling device between the ileum and transverse colon. This is accomplished by excising the corners of each stapled end and stapling the two pieces of bowel together along the antimesenteric border. Make certain that the bowel is not kinked or twisted.

Step 13. Inspect the intraluminal staple line to confirm hemostasis. If necessary, control bleeding with a 3-0 silk gastrointestinal (GI) suture.

Step 14. Close the ileocolotomy using a non-cutting stapling device. Prior to closing, line up the edges of the bowel using Allis clamps to ensure that the stapling device will encompass the entire opening and will reapproximate all layers of the bowel.

Step 15. Place a 3-0 silk GI suture at the distal corner of the anastamosis to relieve any tension.

Step 16. Irrigate the abdominal cavity with 2-4 liters of warm saline. Confirm hemostasis. Confirm laparotomy pad counts, needle counts, and instrument counts.

33

Step 17. Close the abdominal wall with a running #2 nylon suture with interrupted 2-0 polypropylene sutures. Close the skin with staples.

Postop

Intravenous antibiotics are continued for 24 hours. Deep venous thrombosis prophylaxis should be continued until the patient is ambulating. Intravenous fluid is continued until adequate oral intake has been established. The diet should be advanced according to the patient's appetite, lack of nausea and emesis, presence of bowel sounds, and bowel activity. The presence of nausea, hiccups, bloating, and anorexia should alert the physician that the patient should remain NPO. The Foley catheter may be removed on postoperative day 1-3 depending upon the stability of the patient's vital signs and urinary output. The staples may be removed on postoperative day 5-7. This may occur in the clinic, depending upon the pace of the recovery.

Complications

The main complications associated with this procedure include wound infections; pelvic, subphrenic, or subhepatic abscess; anastomotic dehiscence; abdominal wall dehiscence; and prolonged postoperative ileus.

Follow-Up

For benign conditions (adenomatous polyps), accepted screening measures should be followed. For malignancy, the patient should undergo colonoscopy one year following operation and enter lifelong follow-up because of the increased risk of a second colon cancer.

Chapter 34

Hemicolectomy (Right): Laparoscopic

Kenric M. Murayama

Indications

Laparoscopic right colectomy is indicated when removal of the entire right colon is required for benign diseases of the right colon or appendix, such as a large adenomatous polyp which cannot be removed endoscopically. The use of laparoscopic colectomy for frank malignancy of the colon or appendix is controversial because of the risk of port-site metastases, spillage of tumor, and/or inadequate mesenteric resection.

Preop

Standard bowel preparation includes 4 liters of a nonabsorbable electrolyte solution or two bottles of Fleet's phosphosoda to be completed by noon the day prior to surgery and antibiotic preparation to be taken at 1 PM, 2 PM, and 10 PM on the day prior to surgery The choice of antibiotic regimen can be mitronidazole 500 mg po and neomycin 1 g per dose or erythromycin 1 g and neomycin 1 g per dose. Enemas are not routinely administered for right colon surgery. Sequential compression devices are applied to all patients and subcutaneous heparin added for high-risk patients for deep venous thrombosis. Intravenous antibiotic is administered prior to skin incision.

Procedure

Step 1. The patient should be positioned supine with both hands tucked at the sides. A Foley catheter and a naso/orogastric tube are placed in all patients. Standard skin cleansing is performed.

Step 2. If a 5 mm laparoscope is used, then three 5 mm trocars are utilized. If a 10 mm laparoscope is used, then two 5 mm trocars and one 10 mm trocar are used.

Step 3. Access to the peritoneal cavity is obtained for insufflation at the umbilicus. This can be accomplished via either closed (Veress needle) or open (Hasson cannula) technique. If the closed technique is used, then a "saline drop test" should be performed to ensure that the needle is in the peritoneal cavity. Subsequent trocars are placed under direct visualization. One trocar is placed in the epigastric region and the other is placed in the suprapubic region.

Step 4. The abdomen is explored using two atraumatic bowel graspers (Babcock-type) to manipulate bowel. A special effort should be made to examine the surfaces of the liver.

Step 5. The cecum is grasped with the atraumatic bowel clamp placed through the epigastric port and retracted cephalad. The ultrasonic shears placed through the suprapubic trocar are used to mobilize the cecum by dividing the lateral peritoneal attachments starting at the base of the cecum. This dissection is carried cephalad

Northwestern Handbook of Surgical Procedures, edited by Richard H. Bell, Jr. and Dixon B. Kaufman. ©2005 Landes Bioscience.

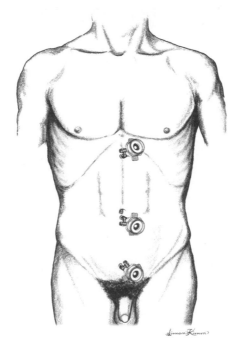

Figure 34.1. Laparoscopic right hemicolectomy. Trocar placement.

along the "white line" of Toldt. The peritoneum overlying the mesentery of the terminal ileum is divided and blunt dissection is used to mobilize the mesentery of the terminal ileum. The terminal ileum should be mobilized for at least 10-15 cm. During this dissection, special effort is made to identify and preserve the right ureter and gonadal vessels.

Step 6. The hepatic flexure is grasped with the atraumatic bowel clamp placed through the suprapubic port and retracted caudad. The ultrasonic shears placed through the epigastric port are used to mobilize the hepatic flexure. The hepatocolic attachments are divided using the ultrasonic shears. Surgical clips may be necessary to control bleeding of large veins. Special care should be taken to avoid injury to the right kidney, duodenum, and right adrenal gland. The right half of the gastrocolic attachments is similarly divided, again taking special care to avoid injury to the duodenum and the middle colic vessels. Adequate mobilization of the hepatic flexure and the right transverse colon is essential to permit extraction of the colon for extracorporeal resection and anastomosis.

Step 7. The right colon is mobilized to the level of the medial border of the right psoas muscle or vena cava. At this point, the mesentery can be divided using laparoscopic linear stapling devices with vascular-sized staples (2.0-2.5 mm). If a laparoscopic linear stapler will be used to divide the mesentery, a 12 mm port will be needed for placement of the stapler into the peritoneal cavity. Alternatively the mesentery can be divided extracorporeally.

Step 8. A 4-5 cm transverse incision is made at the level of the umbilicus at the lateral border of the right rectus muscle. This is carried down through the fascial and muscle layers using a muscle-splitting technique (as in an appendectomy). The peritoneum is opened transversely and the right colon is delivered through the incision.

Step 9. The proximal and distal resection margins are divided using linear staplers in the same fashion as in the open procedure. If the mesentery was not divided intracorporeally, it can be divided in continuity between clamps and the proximal vessels can be ligated.

Step 10. A side-to-side (functional end-to-end) anastomosis is performed using a linear cutting stapler. The open end of the anastomosis can be closed with a linear cutting stapler, a non-cutting stapler, or a handsewn technique. The mesenteric defect should be closed with sutures. The bowel is returned to the peritoneal cavity. Special care should be taken to be certain the bowel or the anastomosis is not traumatized while returning it to the peritoneal cavity.

Step 11. The transverse incision is closed in four layers: peritoneum, muscle layers, anterior fascia, and skin. All layers are irrigated prior to closure to minimize the risk of wound infection.

Step 12. The abdomen is reinsufflated and explored using the laparoscope to examine for hemostasis. The anastomosis is examined as is the mesenteric closure. Any problems can be dealt with laparoscopically.

Step 13. If a 10 mm trocar was used at the umbilicus, the fascia in this incision is closed. Skin incisions are closed with an absorbable, monofilament subcuticular suture closure.

Postop

The Foley catheter and naso/orogastric tubes are removed in the OR. Standard intravenous fluids are administered. Clear liquids are started as soon as the patient passes flatus or has adequate bowel sounds (usually postoperative day 2 or 3). If clear liquids are tolerated, diet is advanced the next day to a general diet and the patient can be discharged.

Complications

Anastomotic leak usually manifests on postoperative days 7-10 by either peritonitis (large leak) or occult fever or failure to progress (small leak with possible abscess). Wound infection is uncommon; if it occurs, it is generally in the transverse incision through which the colon was removed. Intraluminal bleeding from the anastomosis can sometimes be managed conservatively but may require reoperation.

Follow-Up

Follow-up is dictated by the underlying condition for which the colon resection was performed.

34

Chapter 35

Colostomy Closure

John J. Coyle

Indications

Colostomy closure is indicated when the underlying condition which required the colostomy allows it and there is adequate distal colon and rectum to safely re-establish gastrointestinal continuity. Colostomy closure is typically performed 6-12 weeks after creation of the stoma after the inflammation that may have been associated with colostomy placement has resolved.

Preop

The proximal and distal colon should be assessed by colonoscopy, proctoscopy, and/or contrast enema through both colostomy and rectum. Prior to surgery, both a mechanical and an antibiotic bowel prep should be performed. Prophylactic IV antibiotics should be given 30 minutes prior to surgery. A nasogastric tube is inserted.

Procedure

Step 1. An adequate abdominal incision is made for exposure. For closure of a loop colostomy, this can be a horizontal or vertical elliptical incision around the stoma. When the two ends of the bowel are separated, a midline incision or re-opening of the previous laparotomy incision is often required. When closing an end colostomy, the stoma is typically oversewn before prepping the abdomen to minimize spillage.

Step 2. The stoma is separated from surrounding skin, subcutaneous tissue, and fascia until it is possible to pass a finger into the peritoneal cavity circumferentially around the stoma.

Step 3. The proximal and distal colon are mobilized as much as necessary to allow the bowel to be reanastomosed without tension.

Step 4. The colon is anastomosed in a tension-free manner using sutures or stapling instruments. In a divided colostomy, the bowel ends should be resected back to healthy tissue prior to anastomosis. Either a one-layer or two-layer sutured anastomosis can be performed. If the anastomosis is stapled, the two ends of bowel are closed with a linear stapler and then a functional end-to-side anastomosis performed. When closing a loop colostomy, it may be possible to preserve the bridge of tissue connecting the two limbs and close the remainder of the circumference of the bowel. However, this must be done in a way that does not compromise the lumen, and it may be necessary to resect back to fresh ends to effect an adequate anastomosis.

Northwestern Handbook of Surgical Procedures, edited by Richard H. Bell, Jr. and Dixon B. Kaufman. ©2005 Landes Bioscience.

Step 5. The site of the stoma is typically left open and packed to heal by secondary intention. If a separate abdominal incision has been made, it can be closed in layers including skin if contamination has been minimal.

Postop

The nasogastric tube is removed and oral feeding resumed as bowel activity returns. Pain control is switched from parenteral to oral when bowel activity returns.

Complications

Complications of colostomy include wound infection and anastomotic dehiscence with fistula formation or peritonitis.

Follow-Up

The patients should be seen as appropriate in the weeks after colostomy closure to follow healing of the abdominal wound(s). Long-term follow-up is dictated by the underlying condition requiring colostomy.

35

Chapter 36

Colostomy: End Sigmoid with Hartmann's Pouch

David J. Winchester

Indications

Indications for an end sigmoid colostomy include perforated diverticulitis, perforated sigmoid carcinoma, iatrogenic colon perforation following colonoscopy, obstructing unresectable rectal cancer, and rectal injury.

Preop

Most patients who require end sigmoid colostomy with Hartmann's pouch present acutely and therefore are unable to have any preoperative bowel preparation. Intravenous antibiotics should be given upon reaching the diagnosis if peritonitis is suspected or immediately preoperatively if it is not. Broad-spectrum coverage should include gram-negative and anaerobic organisms. Deep venous thrombosis prophylaxis should include subcutaneous heparin or sequential compression devices, as dictated by the patient's risk factors. If possible, the patient should be evaluated by a stomal therapist preoperatively to mark the optimal stomal position. A urinary catheter should be placed.

Procedure

Step 1. The abdomen should be cleansed and draped to expose the area from the pubic symphysis to the nipples. A lower vertical midline incision is made. For obese patients, the incision may need to be extended cephalad. Explore the abdomen and evaluate the pathology. If purulent fluid is present, obtain aerobic and anaerobic cultures.

Step 2. Place a self-retaining retractor such as a Balfour or Omni retractor.

Step 3. Place moist laparotomy pads across the small intestine and pack in the upper abdomen.

Step 4. Incise the peritoneal reflection of the sigmoid colon above and below the site of obstruction or perforation and mobilize the sigmoid to the extent possible. Attempt to identify the left ureter. If the inflammation is severe, maintain a close plane of dissection along the mesenteric border of the sigmoid colon to avoid injuring the ureter, particularly for benign disease. If inflammation or soiling is not severe, attempt to perform a lymphadenectomy for suspected malignancies. For benign disease, there should be no attempt to include the mesentery or lymph nodes with the resection.

Step 5. Identify proximal and distal sites of the sigmoid for resection. The bowel should be as healthy as possible at the resection sites. For unusually severe inflammation or extensive tumor, it may be safest to divide the sigmoid colon

Northwestern Handbook of Surgical Procedures, edited by Richard H. Bell, Jr. and Dixon B. Kaufman. ©2005 Landes Bioscience.

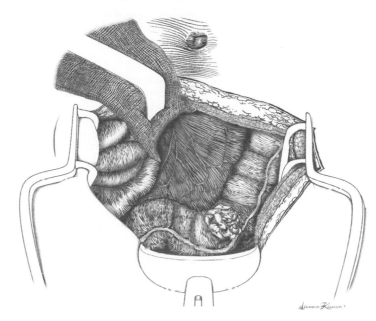

Figure 36.1. End sigmoid colostomy with Hartmann's pouch.

proximal to the site of pathology and leave the pathology in place but diverted until a subsequent operation is performed.

Step 6. Divide the bowel in both locations using an automated stapling device.

Step 7. Clamp, ligate, and divide the mesenteric vessels with 2-0 silk suture.

Step 8. Remove the specimen. Irrigate the abdominal cavity with several liters of warm saline. Inspect for hemostasis.

Step 9. The distal sigmoid colon may be loosely tacked to the sacral promontory or psoas muscle to prevent retraction and facilitate subsequent colostomy reversal providing that adequate length remains after the resection. A long polypropylene suture attached to the Hartmann pouch will also facilitate future identification.

Step 10. Excise a circular piece of skin at the preoperatively marked stomal site. This should be approximately 2 cm in diameter. Make a cruciate incision in the anterior rectus sheath. Do not divide the muscle. Incise the peritoneum. These incisions should allow passage of two fingers.

Step 11. Bring the proximal sigmoid colon out through the stomal incision. The colon should lie freely at the surface of the skin without tension. If this is not possible, mobilize the descending colon prior to closing the abdominal fascia.

Step 12. Place a closed suction drain in the pelvis if an abscess was present. Close the abdominal fascia with a running #2 nylon suture with interrupted 0 polypropylene figure-of-eight sutures. The skin should be loosely approximated and packed with moist gauze in the presence of a perforation. Otherwise, the skin may be closed with staples. Dress the abdominal incision.

Step 13. Excise the staple line on the proximal colon. Sew the edges of the colostomy to the skin with 3 0 Vicryl suture, approximating the dermis to full-thickness purchases of the transected proximal sigmoid. Place a stomal appliance.

Postop

The patient should receive 7-14 days of intravenous antibiotics in the setting of a perforation. Deep venous thrombosis prophylaxis should be continued until the patient is ambulating. Intravenous fluid should be continued until adequate oral intake has been established. The diet should be advanced according to the patient's appetite, lack of nausea and emesis, presence of bowel sounds, and bowel activity. The Foley catheter may be removed on postoperative day 3-5 depending upon the degree of pelvic dissection, the stability of the patient's vital signs, and urinary output.

Complications

Possible procedure-related complications include wound infection, pelvic abscess, abdominal wall dehiscence, prolonged postoperative ileus, and small bowel obstruction.

Follow-Up

Follow-up depends on the nature of the underlying condition. If the colostomy is to be closed, this is typically done 6-12 weeks after the initial procedure.

Chapter 37

Colostomy: Transverse Loop

Richard S. Berk

Indications

Transverse loop colostomy is indicated in the presence of obstruction of the distal (left) colon or rectum where primary resection and anastomosis of the obstructing lesion is unsafe or impossible. Transverse loop colostomy may also be used to protect a low anastomosis of the rectum.

Preop

In obstructed patients, fluid and electrolyte imbalances should be corrected to the extent possible in an expeditious manner prior to surgery. The amount of time available to do this depends on factors such as the degree of dilation of the proximal colon, risk of perforation, etc. In general, it is desirable to verify the presence of mechanical obstruction and the site of obstruction using CT scanning with rectal contrast, water-soluble contrast enema, or endoscopy.

A site for the colostomy should be chosen over the rectus sheath and away from all bony prominences to allow for placement of the ostomy base plate. Preoperative prophylactic antibiotics are given 30 minutes prior to incision and deep venous thrombosis prophylaxis with sequential compression devices or subcutaneous heparin is employed according to the patient's risk factors.

Procedure

Step 1. The patient is positioned supine and general anesthesia used to allow for adequate exploration. Draping exposes the area from xiphoid to umbilicus and just lateral to rectus sheath edges.

Step 2. A midline incision can be made or a transverse incision over the right rectus muscle. If the colostomy is intended to be permanent, placement of the stoma in the left transverse colon through a midline or left rectus incision may help to prevent subsequent prolapse. The incision in any case must be large enough to permit safe manipulation and delivery of the dilated colon segment.

Step 3. The dilated transverse colon is identified by following the omentum down to its colonic attachment and the omentum and dilated segment of colon are delivered through the abdominal wall up into the operative field.

Step 4. The omentum is separated from the colon over a short distance by sharp or cautery dissection through the avascular plane and the omentum then transected at right angles to the colon between clamps, allowing it to remain attached to the colon but fall back into the peritoneal cavity out of the way.

Step 5. If the colon is extremely tense or difficult to handle, it is possible to decompress it with an aspirating needle and close the opening with a suture, incorporating the hole into the site of the eventual colostomy.

Northwestern Handbook of Surgical Procedures, edited by Richard H. Bell, Jr. and Dixon B. Kaufman. ©2005 Landes Bioscience.

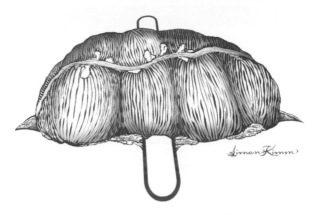

Figure 37.1. Transverse loop colostomy. Supporting rod.

Step 6. A short segment on the mesenteric border of the colon is cleared to allow passage of a small Penrose drain. The Penrose is then used as a handle to pull up the colon segment. It is important to bring enough colon out of the abdominal cavity to eliminate tension once the bridge has been placed under the loop.

Step 7. The fascia is closed with 0 absorbable suture loosely enough to allow for easy passage of a finger between the colon and the fascia. A few of the epiploic fat attachments and/or the colon serosa are tacked to the fascia with Vicryl suture to prevent herniation.

Step 8. The subcutaneous layer is closed with 3-0 absorbable suture and the skin with 3-0 simple sutures. The skin closure should extend close to the colon but be loose enough to allow passage of a finger around the edge of the wound next to the colon.

Step 9. The Penrose drain is withdrawn while simultaneously replacing it with a supporting rod or bridge. The rod is fastened to the skin with nylon suture. The exact nature of fixation to the skin will vary between different devices.

Step 10. The colon loop is opened along one of the tenia colae for approximately one-third of its length, incorporating any previous decompression site. The edges of the opened bowel are everted and gently approximated to the dermis, except at the point where the rod or bridge passes behind the loop, or the opening may be left alone to evert over time.

Step 11. An opening is fashioned in the ostomy disc just large enough to fit around the ostomy flush against the surface, and the disc is placed on the skin.

Postop

The ostomy can be expected to become edematous while the rod is in place. The initial size of the opening in the colon should take this into account to allow for function in spite of the edema. The rod or bridge is left in place 6-10 days and then removed, allowing for shrinkage of the stoma and a reduction in the size of the base plate opening. The wound and surrounding skin should be checked for any signs of infection as part of the follow-up care. A good seal is mandatory each time the bag is changed, although this can be hard to accomplish until the bridge is removed. The absorbable sutures in the derma should be removed at about 10 days if they have not dissolved.

Complications

Bleeding may occur from the stoma site. Necrosis of the stoma may occur if it is brought up under too much tension. Either the proximal or distal end of the stoma may prolapse. Herniation of small bowel through the incision opening may occur if the fascial closure is not snug enough.

Follow-Up

In time, the size of the bag is decreased gradually to fit the smaller stoma. The patient should in general be referred to an enterostomal therapist if they did not see one preoperatively. Closure of the ostomy is planned when the reason for its initiation is no longer in existence. Contrast studies through the stoma and through the rectum are performed prior to colostomy closure.

Chapter 38

Sigmoid Colectomy: Open

Steven J. Stryker

Indications

Indications for sigmoid colon resection include cancer, sessile polyps, diverticular disease, localized Crohn's colitis, volvulus, ischemic colitis, and as an adjunct to suture rectopexy for rectal prolapse.

Preop

When resection of diseased sigmoid is contemplated, it is desirable to rule out significant pathology in the remainder of the colon. In the absence of obstruction or acute inflammation, this is best accomplished by colonoscopy. A mechanical bowel prep includes a clear liquid diet for 48 hours preoperatively, in conjunction with oral phospho-soda 45 cc, given 36 hours and 24 hours preoperatively. Oral antibiotics are of no proven added advantage when perioperative intravenous antibiotics are utilized. Nonetheless, occasionally, three doses of oral neomycin sulfate and metronidazole are given at 18 hours, 17 hours, and 9 hours preoperatively in patients felt to be at high risk for infection. Intravenous ampicillin/sulbactam is given 30 minutes prior to incision. Pneumatic compression sleeves for the lower extremities are used to minimize the risk of deep venous thrombosis. In patients with a prior history of venous thrombosis or pulmonary embolus, subcutaneous heparin may be used as well.

Procedure

Step 1. The patient is placed in dorsal lithotomy position allowing simultaneous access to the abdomen and perineum. The patient is positioned awake to check for comfort in positioning with respect to the back, hips, and knees. Once the patient confirms that the positioning is comfortable, general anesthesia is induced. Care should be taken that there is no excessive pressure on the calves or the lateral aspect of the proximal leg after positioning to avoid peroneal nerve injury and compartmental syndrome postoperatively.

Step 2. The abdomen and perineum are widely prepped with a povidone/alcohol combination prep. The abdomen and perineum are draped to provide wide access to both areas.

Step 3. A lower midline incision is made, taking care to divide the midline fascia down to the pubic symphysis. A thorough abdominal exploration is undertaken to assess the extent of tumor involvement.

Step 4. The lateral and medial peritoneal reflections of the sigmoid colon are incised down past the rectosigmoid junction at the level of the sacral promontory. The left ureter is identified and displaced laterally along with the gonadal vessels.

Northwestern Handbook of Surgical Procedures, edited by Richard H. Bell, Jr. and Dixon B. Kaufman. ©2005 Landes Bioscience.

Step 5. The superior hemorrhoidal vessels and distal sigmoidal vessels are ligated proximally at the base of the mesentery, taking care to identify and avoid the left ureter during this maneuver.

Step 6. The descending colon and proximal rectum are mobilized as necessary to allow a tension-free anastomosis. The mesocolon is divided up to the mesenteric aspect of the proximal margin. The bowel wall at the proximal margin of resection is cleared of pericolic fat for a short distance and a pursestring suture placed on the proximal end as the bowel is transected. The anvil cap of an end-to-end stapling device is inserted into the open proximal end and the pursestring suture is tied.

Step 7. The distal margin of resection is chosen and the mesocolic fat is cleared for a short distance at this point. The bowel is divided with a stapling device in preparation for a double-stapling technique. The specimen is sent for pathologic analysis.

Step 8. The end-to-end stapler is placed transanally and carefully advanced to the stapled rectosigmoid junction. The trocar at the end of the stapler is advanced through the staple line just anterior or just posterior to its midpoint. The trocar is removed and the anvil cap and end of the stapler are connected. The stapler is closed, fired, and slowly withdrawn through the anus.

Step 9. A proctoscope is inserted into the rectum and air is insufflated as the colon is occluded proximal to the anastomosis. The pelvis is filled with saline and the anastomosis is checked for an air leak. If no leak is seen, the rectum is deflated and the saline aspirated from the pelvis.

Step 10. The midline incision is closed in layers.

Postop

Intravenous antibiotics are continued for 24 hours postoperatively. The patient ambulates on the day following surgery. A urinary catheter is left in for 24-48 hours. Clear liquids are begun orally upon resumption of bowel activity.

Complications

Early postoperative complications include: atelectasis, intraabdominal or wound infection, urinary tract infection, postoperative ileus, hemorrhage, venous thrombosis, anastomotic dehiscence, or ureteral injury. Late complications include adhesive small bowel obstruction and anastomotic stricture.

Follow-Up

Patients are checked in the office at 2-4 weeks post-discharge to assess wound healing and overall progress. If the resection has been performed for malignancy, the patient is seen at 3-month intervals for the first 2 years, at 4-month intervals for the 3rd year, and at 6-month intervals during the 4th and 5th years of follow-up.

Chapter 39

Proctocolectomy with Ileal Pouch: Anal Anastomosis

Amy L. Halverson

Indications

Complete removal of the colon and rectum with ileal pouch reconstruction is indicated in individuals with ulcerative colitis who are not responding adequately to medical therapy or who have colonic dysplasia. It is also indicated in individuals with familial adenomatous polyposis (FAP).

Preop

Patients should have had a complete colonoscopy. An enterostomal therapist is helpful for marking an optimal site for the ileostomy and providing preoperative teaching regarding ileostomy care. To allow for a secure fit of the ostomy appliance, the site should be located over the rectus muscle and should be at the apex of any abdominal fold, which occurs with the patient in the sitting position. The site should allow for easy visualization by the patient. Patients should receive mechanical bowel cleansing and intravenous antibiotics 30 minutes prior to operation. Deep vein thrombosis prophylaxis with subcutaneous heparin and/or sequential compression devices should be used based on the patient's risk factors.

Procedure

Step 1. The patient should be in the modified lithotomy position with the hips only slightly flexed. There should be no pressure on the upper or lateral calf. The leg should be supported primarily by the foot and lower calf. Incorrect positioning may result in injury to the peroneal nerve.

Step 2. The abdomen is entered though a midline incision and a complete exploration of the peritoneal cavity performed. The ascending colon is mobilized towards the midline by incising the right peritoneal reflection. Mobilization is continued around the hepatic flexure to the distal transverse colon. The omentum is dissected away from the transverse colon across its entire length and the omentum preserved unless malignancy is suspected. The left colon and splenic flexure are then mobilized.

Step 3. The ileum is divided just proximal to the ileocecal valve. The mesocolon is then divided starting at the point of terminal ileal division and extending to the proximal rectum. If the mesocolon is thickened, it may be prudent to employ suture ligatures on the base of the mesocolon.

Step 4. The upper and mid-rectum is mobilized by dividing the peritoneum on either side of the rectum as it descends into the pelvis. The avascular plane between the rectum and the sacrum is identified and the rectum is mobilized anteriorly

Northwestern Handbook of Surgical Procedures, edited by Richard H. Bell, Jr. and Dixon B. Kaufman. ©2005 Landes Bioscience.

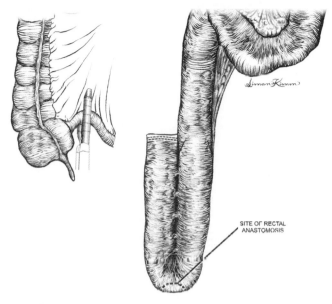

Figure 39.1. Proctocolectomy with ileal pouch.

down to the level of the coccyx. The lateral rectal attachments are divided with electrocautery. The anterior peritoneal reflection is identified where the perirectal fat extends across the midline. The lateral incisions are continued anteriorly in this plane. Dissecting too far anteriorly may result in injury to the vagina in women or bleeding from the seminal vesicles and prostate in men. Circumferential mobilization of the rectum is continued down to the level of the levator muscles. With upward traction on the rectum, the distal rectum is stapled just above the proximal extent of the anal verge.

Step 5. To begin creation of the ileal pouch, the terminal 30 cm of ileum is irrigated with saline.

Step 6. The ileum is then folded into a J-configuration and the two antimesenteric borders of the limbs placed in apposition. The limbs are then stapled along the antimesenteric borders, incorporating both walls with repeated firings of an 8 cm linear stapler. The length of the pouch created is approximately 15 cm.

Step 7. The open end of the J limb is closed with a double-row stapler and imbricated with sutures.

Step 8. An enterotomy is created in the bend at the apex of the pouch and a pursestring suture is placed to hold the anvil of the circular stapler. Prior to inserting the anvil of the stapler, the anterior staple line of the pouch should be checked for bleeding.

Step 9. The end-to-end circular stapler is inserted through the anus. Using the sharp pin, the rectal stump is penetrated from below. The pin is removed and the anvil protruding from the J-pouch attached in its place, and the pouch then stapled to the rectal stump.

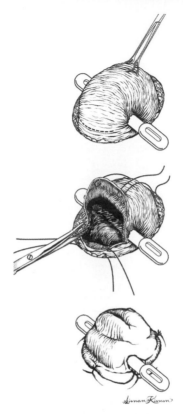

Figure 39.2. Proctocolectomy with ileal pouch. Temporary diverting loop ileostomy.

Step 10. A temporary diverting loop ileostomy may be created by excising a 1.5 cm diameter disc of skin at a preoperatively marked stoma site. The subcutaneous tissue and the anterior rectus sheath are divided using electrocautery. The fibers of the rectus muscle are separated using a blunt curved clamp and the abdomen is entered by incising the posterior rectus sheath and peritoneum.

Step 11. A loop of terminal ileum is brought through the opening in the abdominal wall and a plastic rod for support is placed under the mesenteric edge of the bowel through a small hole in the mesentery immediately adjacent to the bowel wall.

Step 12. A closed suction drain is placed in the pelvis and brought out through a separate stab wound. The abdominal incision is closed in layers

Step 13. A transverse enterotomy is made just above the skin on the distal limb and the bowel everted over the proximal limb. The edges of the opened ileum are sutured to the dermis circumferentially with interrupted 4-0 absorbable sutures.

Postop

The rod under the loop ileostomy may be removed on postoperative day 3. The patient should be examined after 6 weeks to dilate any anastomotic stricture. This can usually be done with gentle digital pressure. The loop ileostomy is closed after approximately 12 weeks. A Gastrograffin enema should be performed to rule out a leak prior to loop ileostomy closure.

39

Complications

Complications of proctocolectomy with ileal pouch-anal anastomosis include anastomotic leak, bleeding from the pouch, pelvic abscess, anal stricture, and wound infection.

Follow-Up

Endoscopic evaluation of the pouch with biopsy of the anal transition zone is performed on an annual or semiannual basis to evaluate for dysplasia. Long-term problems following ileal pouch-anal anastomosis include fistula formation, which may occur in patients with Crohn's disease who were initially thought to have ulcerative colitis or indeterminate colitis. Anal strictures may require dilation, and as many as 20-25% of patients will develop pouchitis.

Chapter 40

Proctocolectomy: Total with Ileostomy

Amy L. Halverson

Indications

Complete removal of the colon and rectum is indicated in patients with Crohn's colitis involving the colon and the rectum. It may also be appropriate for selected individuals with ulcerative colitis or familial adenomatous polyposis (FAP) who are not candidates for reconstruction with an ileal pouch-anal anastomosis. Indications for proctocolectomy in patients with Crohn's disease or ulcerative colitis include symptoms not responding to medical therapy or dysplasia identified on surveillance biopsy.

Preop

Candidates for proctocolectomy should have had a complete colonoscopy. A small bowel contrast study is indicated to look for additional disease in patients with Crohn's disease or to identify small bowel polyps in patients with FAP. The patient should see an eneterostomal therapist for preoperative teaching regarding ileostomy care and to mark a stoma site. Patients should receive mechanical bowel cleansing and intravenous antibiotics 30 minutes prior to operation. Deep vein thrombosis prophylaxis with subcutaneous heparin and/or sequential compression devices should be used based on patient's risk factors.

Procedure

Step 1. The patient should be placed in the modified lithotomy position with hips only slightly flexed. There should be no pressure on the upper or lateral calf. The leg should be supported primarily by the foot and lower calf. Incorrect positioning may result in damage to the peroneal nerve.

Step 2. The abdomen is entered through a midline incision and a complete exploration of the peritoneal cavity performed. The ascending colon is mobilized towards the midline by incising the right peritoneal reflection. Mobilization is continued around the hepatic flexure to the distal transverse colon. The omentum is dissected away from the transverse colon across its entire length and the omentum preserved unless malignancy is suspected. The left colon and splenic flexure are then mobilized. The ileum is divided just proximal to the ileocecal valve. The mesocolon is then divided starting at the point of terminal ileal division and extending to the proximal rectum. If the mesocolon is thickened, it may be prudent to employ suture ligatures on the base of the mesocolon.

Step 3. The rectum is mobilized by dividing the peritoneum on either side of the rectum as it descends into the pelvis. The avascular plane between the rectum and the sacrum is identified and the rectum is mobilized anteriorly down to the level of

Northwestern Handbook of Surgical Procedures, edited by Richard H. Bell, Jr. and Dixon B. Kaufman. ©2005 Landes Bioscience.

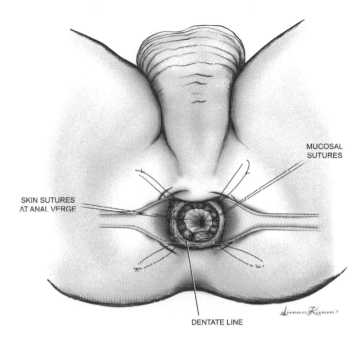

SKIN SUTURES
AT ANAL VERGE

MUCOSAL
SUTURES

DENTATE LINE

Figure 40.1. Total proctocolectomy.

the coccyx. The lateral rectal attachments are divided with electrocautery. The anterior peritoneal reflection is identified where the perirectal fat extends across the midline. The lateral incisions are continued anteriorly in this plane. Dissecting too far anteriorly may result in injury to the vagina in women or bleeding from the seminal vesicles and prostate in men. Circumferential mobilization of the rectum is continued down to the level of the levator muscles.

Step 4. Moving to the perineum, exposure is obtained by effacing the anal canal with circumferential sutures tied to the skin of the buttocks.

Step 5. Submucosal injection of saline with epinephrine facilitates dissecting the mucosa off of the underlying internal sphincter muscle. Transanal mucosectomy is begun with a circumferential incision at the bottom of the mucosa and working proximally. No mucosa should be left behind.

Step 6. After completely removing the mucosa and obtaining hemostasis, the effacing sutures are removed and the anal canal is closed with 2 to 3 rows of absorbable suture. The skin is closed with 3-0 absorbable interrupted vertical mattress sutures.

Step 7. At the site chosen for the ileostomy, a 1.5 cm diameter disc of skin is excised. The subcutaneous tissue and the anterior rectus sheath are divided using electrocautery. The fibers of the rectus muscle are separated using a blunt curved clamp and the posterior rectus sheath and peritoneum opened.

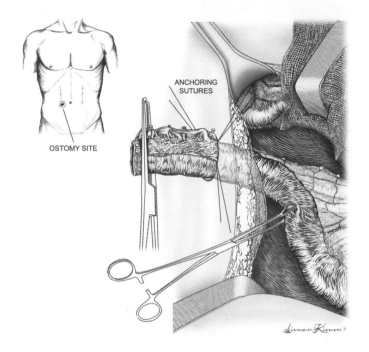

Figure 40.2. Total proctocolectomy. Ileostomy.

Step 8. The cut end of the ileum is then brought through the defect in the abdominal wall. The ileal mesentery is secured to the subcutaneous tissue with interrupted sutures at each border between serosa and mesentery and at the cut edge of the mesentery.

Step 9. The cut edge of mesentery is tacked to the anterior abdominal wall. A closed suction drain is placed in the pelvis.

Step 10. After closure of the abdominal wound, the end of the ileum is everted and secured to the dermis with circumferential interrupted absorbable sutures.

Postop

Once the ileostomy functions, the diet is advanced as tolerated. Ileostomy care teaching should be continued in the postoperative period.

Complications

Bleeding may occur internally or from the stoma. Pelvic infection, perineal wound dehiscence, or stomal necrosis/retraction are other potential complications.

Follow-Up

Patients with Crohn's disease should be followed indefinitely, since they may be at risk for developing small bowel Crohn's disease. Patients with FAP may develop polyps elsewhere in the gastrointestinal tract.

Chapter 41

Anal Fistulotomy

Joseph P. Muldoon

Indications
Anal fistula

Preop
No preparation is required other than routine preoperative evaluation. Regional anesthesia is preferred, but general anesthesia and monitored local anesthesia with sedation are also acceptable choices.

Procedure
Step 1. The patient is placed in the prone jackknife position, with the buttocks separated using tape. The perianal area is prepped with povidone-iodine.

Step 2. The external opening of the fistula is identified on the perianal skin and a probe passed to assess direction of the tract into the rectum.

Step 3. An anoscopic evaluation of the anal canal and lower rectum is performed, and the internal opening of the fistula is searched for. If the opening is not obvious, dilute hydrogen peroxide is injected through the external opening using a 10 cc syringe and an 18-gauge angiocath while the anoscope is in place. Bubbles will mark the site of the internal opening.

Step 4. The extent of incorporated sphincter muscle is assessed by palpation.

Step 5. If less than 50% of sphincter muscle will be incorporated by a fistulotomy, the metal probe is then passed through the length of the fistula tract and the overlying tissues divided down to the probe. If greater than 50% of sphincter muscle is at risk of division, the distal (epidermal) end of the fistulous tract is opened and a vessel loop placed around the sphincter as a seton, and the muscle is not divided at this time.

Step 6. The opened tract is curetted to remove granulation tissue. Hemostasis is obtained using electrocautery.

Postop
Anal fistulotomy is an outpatient procedure. The patient should be maintained on stool softeners and given pain medication. Sitz baths are started the following day.

Complications
Complications include bleeding, perianal abscess, recurrent fistula, and fecal incontinence.

Northwestern Handbook of Surgical Procedures, edited by Richard H. Bell, Jr. and Dixon B. Kaufman. ©2005 Landes Bioscience.

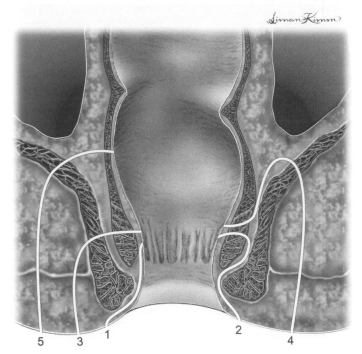

1 SUBCUTANEOUS
2 INTERSPHINCTERIC
3 TRANSSPHINCTERIC
4 SUPRALEVATOR
5 EXTRASPHINCTERIC

Figure 41.1. Anal fistulotomy. Common locations of anal fistulae.

Follow-Up

Patients should be seen approximately one week after the procedure and then followed as appropriate until healing occurs. Fecal continence should be confirmed by history in the postoperative period.

41

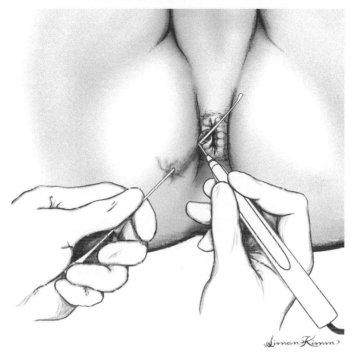

Figure 41.2. Anal fistulotomy.

Chapter 42

Anal Fissure: Lateral Internal Sphincterotomy

Joseph P. Muldoon

Indications

Lateral internal sphincterotomy is indicated for the treatment of symptomatic chronic anal fissure.

Preop

The patient should be on a liquid diet for 24 hours prior to surgery and if possible take a disposable enema 2 hours before surgery. Anesthetic choices are local with intravenous sedation (preferred), regional, or general.

Procedure

Step 1. The patient is placed in the prone jackknife position, buttocks separated using tape. The perianal area is prepped with povidone-iodine.

Step 2. An anoscopic evaluation of the anal canal is performed.

Step 3. The intersphincteric groove is identified by palpation on the right side of the anal canal.

Step 4. A longitudinal incision is made in the mucosa overlying the intersphincteric groove.

Step 5. The intersphincteric plane is developed bluntly with a hemostat.

Step 6. The internal sphincter is divided. The incision is started distally and carried proximally to the level of the dentate line.

Step 7. The wound is irrigated with saline and hemostasis assessed.

Step 8. The mucosal defect is closed with a 3-0 chromic suture.

Postop

This is an outpatient procedure. The patient should be maintained on stool softeners, and sitz baths should be started the following day.

Complications

Complications include bleeding, perianal abscess or fistula, and fecal incontinence.

Follow-Up

Patients should be seen in one week and then regularly until healing occurs. Fecal continence should be confirmed by history in the postoperative period.

Northwestern Handbook of Surgical Procedures, edited by Richard H. Bell, Jr. and Dixon B. Kaufman. ©2005 Landes Bioscience.

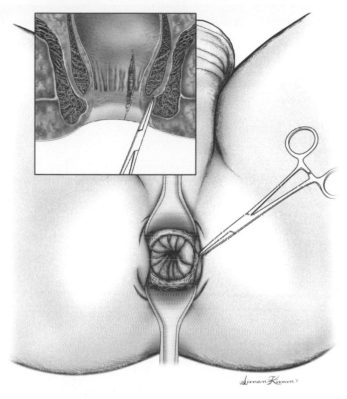

Figure 42.1. Anal fissure. Lateral internal sphincterotomy.

Chapter 43

Anorectal Abscess: Drainage Procedure

Amy L. Halverson

Indications
Acute perirectal abscess.

Preop
This procedure may be performed using local anesthesia with or without intravenous sedation, although some patients may require general anesthesia. A digital rectal examination may be too painful and is not necessary if the site of the abscess is readily apparent.

Procedure
Step 1. Identify area of maximal erythema, induration, and/or fluctuance.
Step 2. Prep skin and infiltrate with local anesthetic.
Step 3. Incise the area identified in Step 1 with a 1-2 cm incision placed over the abscess at the point closest to the anal verge.
Step 4. Allow drainage of purulent fluid. This may be facilitated with gentle insertion of a curved clamp.
Step 5. Insert a 10-14 F mushroom catheter into the abscess cavity and trim drain.

Postop
Consider systemic antibiotics for patients with fever, cellulitis, diabetes, or those who are otherwise immunocompromised.

Complications
Incomplete drainage may result in progression of infection. Inadvertent division of anal sphincter muscles may cause anal incontinence. If symptoms do not resolve after drainage consider examination under anesthesia or imaging (CT or MRI) to identify an undrained collection.

Follow-Up
Examine the patient 2-4 weeks after procedure. Remove drain and allow wound to heal secondarily. Persistent drainage may require examination under anesthesia to identify a fistula tract.

Northwestern Handbook of Surgical Procedures, edited by Richard H. Bell, Jr. and Dixon B. Kaufman. ©2005 Landes Bioscience.

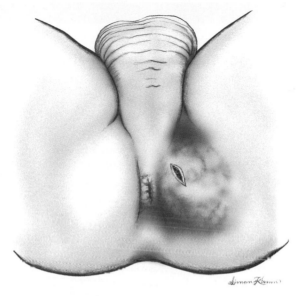

Figure 43.1. Anorectal abscess drainage.

Chapter 44

Internal Hemorrhoids: Band Ligation

Steven J. Stryker

Indications

Rubber band ligation of internal hemorrhoids is indicated for bleeding or prolapse. It is most effective for Grade 1 and 2 internal hemorrhoids and less effective for Grade 3 and 4. Band ligation is contraindicated for the treatment of external hemorrhoids.

Preop

A detailed history is taken to confirm that the internal hemorrhoids are symptomatic. In the case of bleeding, examination of the colon and rectum is performed as appropriate to exclude other pathology. Immediately prior to surgery, a disposable enema is administered to evacuate the rectum.

Step 1. Sedation and local anesthesia are not necessary. The patient is positioned in the prone-jackknife position. Two bands are loaded on the ligating instrument using the cone-shaped loader.

Step 2. An anoscope is inserted and the internal hemorrhoidal quadrants are identified.

Step 3. In general, it is best to perform band ligation of only one quadrant at a session, typically the largest hemorrhoid. Less often, ligation of two or even three quadrants can be accomplished.

Step 4. The grasper is passed through the ligating instrument, and the columnar mucosa just proximal to the targeted hemorrhoid is grasped. This mucosa is drawn well into the ligating instrument and the instrument is deployed, firing the band across this tissue. Great care must be taken to avoid inadvertently capturing any anoderm within the band to minimize postprocedural discomfort.

Step 5. The strangulated tissue is inspected and should be seen just proximal to the hemorrhoid. The blood supply to the targeted hemorrhoid is theoretically interrupted and the hemorrhoid should regress over the ensuing days and weeks.

Postop

The patient is asked to limit physical activity the day of the procedure, but can resume normal activities the following day. Bulk-forming laxatives and nonnarcotic analgesics are recommended.

Complications

Immediate, severe pain can occur and is a sign that the band has been placed too distal in the anal canal. This usually requires the removal of the band by carefully incising it. Postoperative sepsis is a rare, but life-threatening complication, and use

---

Northwestern Handbook of Surgical Procedures, edited by Richard H. Bell, Jr. and Dixon B. Kaufman. ©2005 Landes Bioscience.

of band ligation in immunocompromised individuals is strongly discouraged. Signs of post-banding sepsis include delayed anorectal pain, urinary retention, and fever. Finally, delayed hemorrhage can occur in the first 7-10 days.

Follow-Up

The patient is seen 4-8 weeks after band ligation and the adequacy of the initial treatment determined. Additional band ligation can be performed at this time, if necessary.

44

Chapter 45

Inguinal Hernia Repair with Mesh: Open

Ermilo Barrera, Jr.

Indications

Open repair is indicated for primary or recurrent inguinal hernia in patients who are suitable operative risks.

Preop

General, spinal, epidural, or monitored local anesthesia with sedation may be chosen as anesthetic techniques and the choice should be discussed with the patient. Intravenous prophylactic antibiotic is given 30 minutes prior to the skin incision. Deep vein thrombosis prophylaxis should be used in patients with risk factors for thromboembolism.

Procedure

Step 1. The patient is placed in the supine position. An oblique incision over the inguinal canal is made, using the pubic tubercle as a guide for the medial end of the incision.

Step 2. After opening the external oblique fascia, the ilioinguinal nerve is identified and preserved.

Step 3. Blunt or sharp dissection and a finger are used to surround and isolate the spermatic cord at the pubic tubercle. A Penrose drain is placed around the cord. During this dissection, the genitofemoral nerve should be identified and protected.

Step 4. The hernia sac is identified and separated completely from cord structures back to the level of the internal inguinal ring. In the case of a direct inguinal hernia, the internal ring should be examined to exclude the possibility of an additional indirect sac.

Step 5. For an indirect hernia, the sac is ordinarily treated by high ligation and excision of the sac or inversion into the internal inguinal ring. If the hernia is a sliding hernia, the sac can be inverted back into the internal ring. For direct hernias, the sac is inverted into the fascial defect. If desired, a mesh plug can be used to maintain reduction of the sac by placing it over the sac and securing it to the circumference of the defect

Step 6. Place an onlay patch of mesh over the inguinal canal. The spermatic cord is brought through a "key hole." The mesh should be secured with sutures or staples medially at the pubic tubercle, laterally into muscle beyond the external ring, superiorly to the conjoint tendon, and inferiorly to the shelving edge of the inguinal ligament.

Step 7. The external oblique fascia is closed, taking care not to make the external ring too tight, which can cause venous outflow obstruction from the testicle. The skin is closed with subcuticular suture.

Northwestern Handbook of Surgical Procedures, edited by Richard H. Bell, Jr. and Dixon B. Kaufman. ©2005 Landes Bioscience.

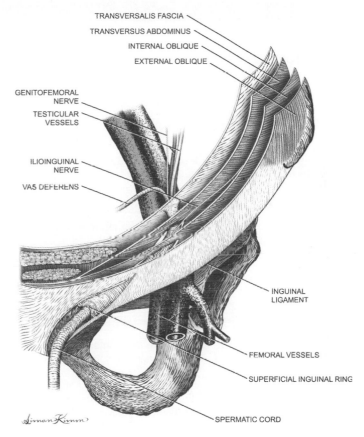

TRANSVERSALIS FASCIA
TRANSVERSUS ABDOMINUS
INTERNAL OBLIQUE
EXTERNAL OBLIQUE

GENITOFEMORAL NERVE
TESTICULAR VESSELS
ILIOINGUINAL NERVE
VAS DEFERENS

INGUINAL LIGAMENT
FEMORAL VESSELS
SUPERFICIAL INGUINAL RING
SPERMATIC CORD

Figure 45.1. Inguinal hernia repair. Anatomy.

Postop

Patients can be discharged within a few hours after surgery if stable. Normal activities can generally resume in 2-4 weeks, although strenuous activity and heavy lifting are generally avoided for about 4-6 weeks.

Complications

Injury to the ilioinguinal nerve or genitofemoral nerve can result in chronic groin pain. Hematoma or seroma may occur. Infections may occur in the wound. If the mesh becomes exposed or infected, it may need to be removed. Mesh can migrate or erode. Recurrence occurs in 1-2% of patients operated for the first time.

Follow-Up

Patients may be seen periodically until they return to full activity.

Chapter 46

Inguinal Hernia Laparoscopic Repair: Extraperitoneal Approach

Mark Toyama

Indications

Laparoscopic inguinal hernia repair is particularly indicated for recurrent or bilateral hernias. Its role in the management of first-time unilateral hernias is debatable.

Preop

Preoperative prophylactic antibiotic should be given intravenously 30 minutes prior to skin incision.

Procedure

Step 1. The patient is placed in the supine position with arms tucked at the sides. A Foley catheter is inserted into the bladder. The surgeon stands on the side opposite the hernia. Monitor(s) are placed at the foot of the table. The skin incision is placed just inferior to the umbilicus and dissection is carried down to the rectus sheath.

Step 2. A small incision is made in the anterior rectus sheath.

Step 3. The rectus abdominis muscle is bluntly dissected to expose the posterior rectus sheath.

Step 4. Blunt dissection is done to develop the space between the back side of the rectus muscle and the peritoneum. A finger or small retractor works well for this.

Step 5. A balloon dissector is then placed into the preperitoneal space and carefully advanced inferiorly to the level of the symphysis pubis. The balloon is inflated under direct vision of the laparoscope, creating a working area in the preperitoneal space.

Step 6. The balloon is removed and a 10 mm or 12 mm blunt trocar is placed into the preperitoneal space and secured. The preperitoneal space is insufflated with CO_2.

Step 7. A 30° laparoscope is placed into the preperitoneal space.

Step 8. Additional ports are placed. A combination of 2 mm and 5 mm ports can be use in a variety of configurations. The ports should be placed under direct vision, taking care to avoid puncturing the thin peritoneum.

Step 9. The preperitoneal space is bluntly dissected, reducing indirect, direct, or femoral hernias back into the peritoneal cavity.

Step 10. Very large indirect hernia sacs can be divided and the proximal end secured with a ligature, leaving the distal portion of the sac open and in situ. Care is taken to preserve all spermatic cord structures in men.

Northwestern Handbook of Surgical Procedures, edited by Richard H. Bell, Jr. and Dixon B. Kaufman. ©2005 Landes Bioscience.

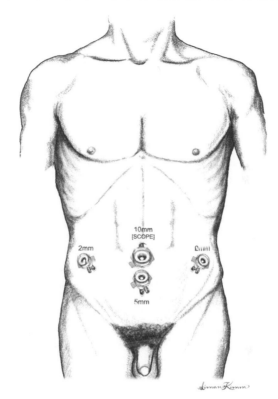

46

Figure 46.1. Laparoscopic inguinal hernia repair. Trocar placement.

Step 11. A large piece of mesh is then placed into the preperitoneal space and oriented to cover the direct, indirect, and femoral spaces.

Step 12. The mesh is secured with a tacking or stapling device to prevent mesh migration. The number of tacks required is variable, but this is done in such a manner as to avoid injuring or incorporating the ilioinguinal, iliohypogastric, lateral femorocutaneous, or genitofemoral nerves as well as any vascular or cord structures.

Step 13. After the mesh is placed and secured, the preperitoneum is desufflated under direct vision to ensure that the mesh remains flat and in the appropriate position.

Step 14. The skin incisions are closed with subcuticular sutures.

Postop

Postoperative management is similar to that of open hernia repair. The Foley catheter is removed before the patient leaves the operating room. Patients are discharged home when they can tolerate oral intake and void.

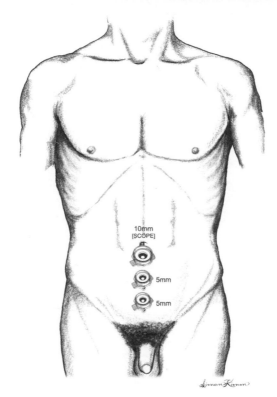

Figure 46.2. Laparoscopic inguinal hernia repair. Trocar placement.

Complications

A number of injuries are possible during laparoscopic preperitoneal hernia repair. These include nerve injury, vascular injury, bladder injury, colon or small bowel injury, testicular devascularization, and vas deferens injury. Urinary retention and/or infection may occur.

Seromas or hematomas may form in the dissected preperitoneal space. Pubic/pelvic osteitis may occur.

Wound infections are relatively rare. Mesh complications include infection, migration, and erosion. Finally, there is about a 2-5% chance of hernia recurrence.

Follow-Up

Patients are followed in the office approximately 2 weeks after their operation. Patients are instructed to avoid heavy lifting and straining for approximately 4-6 weeks after the operation.

46

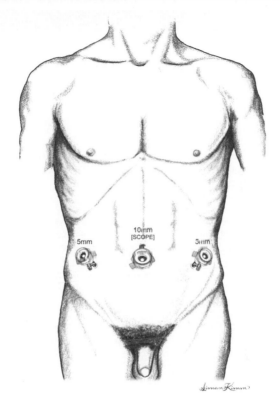

Figure 46.3. Laparoscopic inguinal hernia repair. Trocar placement.

Chapter 47

Ventral Hernia Repair: Open

John J. Coyle

Indications

In general, ventral hernias should be repaired in patients who are good operative risks to avoid the possibility of strangulation. Repair is definitely indicated in the presence of symptoms (pain, nausea, vomiting, etc.) or if the hernia cannot be reduced.

Preop

Patients should be given a preoperative systemic antibiotic for wound infection prophylaxis 30 minutes before operation. In cases where there is likely to be colon in the hernia sac or adhesions to the colon, a mechanical and pharmacologic bowel prep is indicated. Patients should be treated with sequential compression devices or subcutaneous heparin according to their preoperative risk factors for thromboembolism. Most procedures should be done under general anesthesia.

Procedure

Step 1. A skin incision should be made that will expose the full length of the hernia defect. In cases where a hernia has occurred in a portion of a previous incision, it is important to have adequate exposure to examine the remainder of the previous incision for possible defects.

Step 2. The subcutaneous tissue is dissected down to the hernia sac, at which point subutaneous flaps are raised all around the hernia until normal fascia can be identified on all sides of the hernia. The peritoneal sac is then opened. Often, there is redundant peritoneum that can be excised back to expose the fascial edges of the defect. It is critical to definitively identify fascia at all margins of the defect.

Step 3. An assessment is made of the amount of tension that would be created with a primary repair of the defect. In general, defects more than 2-3 cm in diameter will not be amenable to primary closure.

Step 4. If primary closure is entertained, relaxation incisions can be made in the anterior rectus sheath or other anterior layers of abdominal fascia to decrease closure tension.

Step 5. Primary closure can be performed with direct approximation or a pants-over-vest method.

Step 6. If mesh is to be placed, the mesh should be cut in the shape of the defect but about 1-2 cm larger in all diameters. The mesh is sewn to the underside of the fascial edges of the defect using interrupted vertical mattress sutures of a monofilament suture.

Northwestern Handbook of Surgical Procedures, edited by Richard H. Bell, Jr. and Dixon B. Kaufman. ©2005 Landes Bioscience.

Step 7. If the subcutaneous fat can be closed over the repair without tension, this should be done with absorbable sutures. Often, this is not possible, If there will be a subcutaneous cavity over the repair, it is best to place a closed suction drain over the repair and bring it out through a separate stab wound in the skin.

Step 8. The skin is closed with staples or suture.

Postop

If an extensive lysis of adhesions has been performed, it may be appropriate to leave a nasogastric tube in place until bowel activity has returned. Ordinarily, however, early feeding can be initiated. The patient should be able to ambulate on the day following surgery in most cases.

Complications

The most important early complication of ventral hernia repair is wound infection, which can present a major problem if mesh is exposed or involved. The most significant late complication is hernia recurrence.

Follow-Up

The patient should be followed until healing is complete and normal activity resumed. The patient should be instructed about the risk and signs of recurrence and asked to return as needed should symptoms develop.

Chapter 48

Ventral Hernia Repair: Laparoscopic

Kenric M. Murayama

Indications

In general, ventral hernias should be repaired in patients who are good operative risks to avoid the possibility of strangulation. Repair is definitely indicated in the presence of symptoms (pain, nausea, vomiting, etc.) or if the hernia cannot be reduced.

Preop

The alternative of open ventral hernia repair should be discussed with the patient. If the patient chooses laparoscopic surgery, a careful review of previous operations and examination of the abdomen is carried out to help plan potential access sites. On the day of surgery, patients should be given a preoperative systemic antibiotic for wound infection prophylaxis 30 minutes before operation. In cases where there is likely to be colon in the hernia sac or adhesions to the colon, a mechanical and pharmacologic bowel prep prior to surgery is indicated. Patients should be treated with sequential compression devices or subcutaneous heparin according to their preoperative risk factors for thromboembolism. A Foley catheter and nasogastric tube are placed immediately after the induction of anesthesia.

Procedure

Step 1. The entire abdomen is prepped and draped in sterile fashion. A sterile plastic barrier is utilized to avoid contact of the prosthetic material with exposed skin.

Step 2. Access is first obtained away from prior surgical sites, on the side opposite previously dissected areas. For example, if a patient has had a low anterior resection and has an incisional hernia, access should first be obtained on the right side of the abdomen to avoid placement of the initial operating port through adhesions.

Either a Veress needle or open technique can be used for initial access to the peritoneal cavity. Veress needle access can be difficult away from the midline. If Veress needle access is initially unsuccessful, the surgeon should have a low threshold for converting to an open access technique (e.g., Hasson cannula).

Step 3. An angled laparoscope is used to permit the surgeon to see around the edges of adhesions. The abdomen is explored and adhesions are assessed. Sites are selected for subsequent port placement. In general, two ports are placed on the same side as the first trocar and at least one other port is placed on the contralateral side to facilitate the later securing of the mesh.

Northwestern Handbook of Surgical Procedures, edited by Richard H. Bell, Jr. and Dixon B. Kaufman. ©2005 Landes Bioscience.

Step 4. Adhesions are divided using either sharp dissection with electrosurgical cautery (staying away from bowel) or ultrasonic shears. Traction is the key to facilitate division of adhesions, and using two hands to dissect helps in manipulation of bowel and tissues. Occasionally, initial adhesiolysis must be done through one port to "clear" space for placement of subsequent ports. Special care must be taken to avoid injury to bowel.

Step 5. Once all hernia contents have been reduced and the edges of the defect are well exposed, the defect is transilluminated from the abdomen and the defect margins are marked with a pen on the plastic barrier drape.

Step 6. An appropriately sized piece of polytetrafluoroethylene (PTFE) mesh is selected. It must overlap the edges of the defect by at least 2-3 cm circumferentially when the abdomen is insufflated. When the mesh is laid on the plastic barrier drape ("non-adhesion-forming" side of the mesh against the plastic barrier drape), the previously made ink marks identifying the defect edges are transferred to the prosthetic material. This aids in trimming of the mesh to 2-3 cm beyond the edges of the defect.

Step 7. At least four, but preferably six or eight, nonabsorbable stay sutures are placed circumferentially around the edges of the mesh, spaced equidistantly. The sites of suture placement are marked on the abdominal wall for future passage of the sutures.

Step 8. The mesh with the sutures is passed through the largest port (generally a 12 mm port except for the smallest mesh which can be placed through a 10 mm port) by rolling the mesh as tightly as possible.

Step 9 The mesh is oriented properly and unfurled in this orientation.

Step 10. The suture passer (disposable or reusable) is passed through the previously marked skin sites and each of the suture ends is grasped. For each suture site, the suture passer must be passed twice through the same skin puncture, but different fascial sites (1 cm apart) so that the suture ends can be tied external to the fascia.

Step 11. Once this is completed, the mesh is secured circumferentially using a laparoscopic tacking device. Tacks are placed 1-1.5 cm apart. Special care should be taken to avoid plication of the mesh.

Step 12. The abdomen is inspected for hemostasis, and any bowel that was dissected is examined for leakage or injury. If there are no problems, all ports are removed under direct visualization to assure that there is no port site bleeding.

Step 13. The fascia is closed at port sites larger than 5 mm. Skin is closed at all port sites with absorbable subcuticular suture and/or sterile tapes. Drains are not used.

Postop

In general, pain is fairly significant in the first 24-48 hours and ileus is not uncommon. Patients are generally hospitalized for one or two nights. Seroma formation in the previous hernia soft tissue defect is common. While this can be alarming to the patient, nothing should be done unless the seroma is symptomatic or signs of infection appear. In general, seromas and hematomas will resolve in 3-4 weeks. If drainage is required, this can be done percutaneously, but should be avoided if possible. Vigorous physical activity should be limited for 2 weeks while tissue ingrowth occurs, but there is no limitation necessary thereafter.

Complications

Occult bowel injury is a serious potential complication. Patients who do not seem to be recovering appropriately within 24-48 hours or who demonstrate signs of peritonitis (fever, elevated white blood cell count) should have an abdominal CT scan and possible urgent return to surgery.

Follow-Up

The patient should be followed until healing is complete and normal activity resumed. The patient should be instructed about the risk and signs of recurrence and asked to return as needed should symptoms develop.

48

Chapter 49

Exploratory Laparotomy: Open

Michael A. West

Indications

Open exploratory laparotomy is indicated in settings where a surgically correctable problem may exist in the abdomen. The most common indications for open exploratory laparotomy include conditions of acute intraabdominal infection and acute traumatic injuries. Open exploration is particularly useful when questions arise concerning the integrity or the condition of the bowel. Whereas CT can provide very accurate anatomic information regarding retroperitoneal and solid organ structures it is much less reliable for evaluation of the bowel. Diagnostic laparoscopy may be considered because it is less invasive; however, it also has lower diagnostic accuracy for evaluating the intestine. An advantage of open laparotomy is the ability to address the primary problem, whatever it might be.

Preop

Prior to performing exploratory laparotomy the patient should have appropriate venous access and should (if possible) be well resuscitated. It is advantageous to place a Foley catheter prior to abdominal exploration. When performing exploratory laparotomy for blunt trauma have adequate operative suction (two suctions), lighting, and carefully position the patient such that the chest and/or mediastinum can be accessed intraoperatively. Antibiotic prophylaxis should be instituted prior to making the incision. Choice of agent should be based on the pathogens likely to be encountered. Second-generation cephalosporins or other agents that cover aerobic and anaerobic enteric pathogens are frequently used. General endotracheal anesthesia is required, along with good muscle relaxation.

Procedure

Step 1. The patient is placed in the supine position. A midline abdominal incision is made from the xiphoid to the pubis. When rapid abdominal access is required in traumatic situations the incision can be made most rapidly with 2-3 scalpel passes. The first pass cuts through the subcutaneous tissue down to the level of the fascia. A second pass of the scalpel can be used to incise the fascia in the midline. The peritoneum can then be entered using a scissors. It is best to complete the fascial incision prior to incising the peritoneal cavity as any tamponade-effect will be released once the peritoneal cavity is entered.

Step 2. On entering the abdominal cavity, pay attention to where bleeding or contamination appears to be arising. It is best not to be distracted by bowel injury/contamination in the setting of massive hemoperitoneum. The peritoneal cavity is packed with laparotomy pads in the four quadrants of the abdomen, but packing should be done first in the quadrant that is most likely to be the source of the

Northwestern Handbook of Surgical Procedures, edited by Richard H. Bell, Jr. and Dixon B. Kaufman. ©2005 Landes Bioscience.

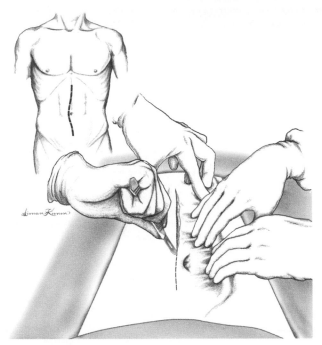

49

Figure 49.1. Exploratory laparoscopy.

bleeding. Bowel injuries with ongoing enteric leakage can be controlled with temporary mass ligatures or application of noncrushing clamps.

Step 3. If there is massive bleeding, temporary control of hemorrhage can be obtained with compression of the aorta at the diaphragmatic hiatus. This can be performed using digital pressure, pressure from a Richardson retractor (back of one blade), or with an aortic occluder.

Step 4. Once hemorrhage is stabilized, the surgeon should explore the four corners of the abdomen while removing the temporary packs (if applicable). It is important to utilize a systematic approach to ensure that all intraabdominal structures are visualized. Particular care should be taken to ensure that relatively inaccessible areas (diaphragm, lesser sac, pelvis) are carefully evaluated. Warm saline irrigation of the relevant quadrant should be performed while the exposure is optimized.

Step 5. Exploration is begun in the left upper quadrant. It is important to visualize the diaphragm, spleen, stomach, and gastroesophogeal junction. Appropriate control measures or repacking should be instituted as applicable if specific injuries are identified.

Step 6. Attention is next directed to the right upper quadrant, taking care to visualize the diaphragm, the diaphragmatic surface of the liver, the integrity and condition of the gallbladder, the lateral aspect of the liver, and the undersurface of the liver.

Step 7. The right lower quadrant of the abdomen is visualized next, paying particular attention for the presence of any bowel or bladder perforation or retroperitoneal hematoma in the area of the iliac vessels.

Step 8. In examining the left lower quadrant, attention is directed to assessing the integrity and condition of the sigmoid colon and looking for evidence of retroperitoneal injury.

Step 9. The integrity of the small bowel is next determined by "running" the small bowel from the ligament of Treitz to the ileocecal valve. The surgeon should make a mental note as to whether there is any evidence of a central retroperitoneal hematoma. Both sides of the small bowel should be examined. The mesentery is inspected simultaneously.

Step 10. The colon is inspected beginning at the cecum with evaluation of the appendix and periappendiceal structures. Inspection continues by examining the cecum and continuing up the right colon carefully examining the hepatic flexure. Examination continues by assessing the transverse colon with the omentum reflected cephalad. Complete evaluation of the splenic flexure may require division of the splenocolic ligament. Evaluation continues by inspecting the left colon and sigmoid colon. Complete evaluation of the right, left, or sigmoid colon may require mobilization of these structures by division of the lateral peritoneal attachments (white line of Toldt) and medial reflection.

Step 11. The anterior surface of the stomach and duodenum should be examined next. In the process, the surgeon should pay particular attention to whether there is any evidence of blood or inflammation in the lesser sac by closely examining the lesser omentum.

Step 12. The pancreas and lesser sac are evaluated next by entering the lesser sac. This is accomplished by making an incision on the undersurface of the omentum just cephalad to the transverse colon. This is most easily accomplished to the left side of the midline. Both the anterior surface of the pancreas as well as the posterior aspect of the stomach can be inspected through this incision.

Step 13. If duodenal injury is suspected the duodenum can be mobilized by performing a Kocher maneuver (lateral incision of retroperitoneum and medial reflection of the duodenum). This also allows visualization of the right renal vein and inferior vena cava. The third portion of the duodenum can be visualized by performing a Cattel maneuver (division of the lateral attachment of the cecum and medio-cephalad cecal reflection).

Step 14. If bowel or solid organ injuries are identified they should be addressed prior to closure.

Step 15. At the completion of the procedure the abdomen should be irrigated with warm saline solution (antibiotics are not required in this fluid).

Step 16. The fascia is closed with running 0-monofilament sutures beginning at the superior and inferior aspects of the incision and meeting in the middle. This skin is either closed or left open as dictated by the intraoperative findings.

Postop

Careful postoperative management and evaluation of the fluid balance should be performed. In instances of trauma or infection that require significant resuscitation, abdominal compartment syndrome can occur with severe hemodynamic and metabolic consequences. The surgical incision should be examined on a daily basis and opened if there is evidence of infection present.

Complications

The most common complication of laparotomy is wound infection. Inadequate exploration can result in missed injuries. Wound dehiscence can also occur. Hypothermic coagulopathy can complicate prolonged exploratory laparotomy in many patients. Abdominal compartment syndrome can occur.

Follow-Up

The patient should be followed until wounds are healed. Long-term follow-up depends on the nature of the underlying disease/injury.

49

Section 2: Endocrine
Section Editor: Richard H. Bell, Jr.

Chapter 50

Adrenalectomy: Laparoscopic

Peter Angelos

Indications

Laparoscopic adrenalectomy is indicated for the removal of functional adrenal tumors or nonfunctional tumors that have met appropriate size criteria.

Preop

Whenever operating on an adrenal gland, it is essential that a pheochromocytoma has been adequately ruled in or out. This is best done with a 24-hour urine sample for vanillylmandelic acid (VMA), catecholamines, and metanephrines. If the patient does have a pheochromocytoma, preoperative alpha-adrenergic blockade for a period of 2-4 weeks and rehydration are necessary. If an aldosterone-secreting tumor is the cause for the surgery, the patient's potassium level should be carefully monitored and normalized preoperatively. All patients are given a mechanical bowel prep the day before surgery.

Procedure

Step 1. The operating room is set up with the monitors just off the patient's shoulders. After a general endotracheal anesthetic has been given, the patient is placed in the lateral decubitus position with the side of the tumor up. The patient is placed on the operating table in such a way that the kidney rest can be elevated and the table flexed, maximizing the space between the costal margin and the anterior superior iliac spine. The surgeon stands facing the patient's abdomen.

Step 2. The patient's entire side extending down the abdomen and the back is prepped and draped in the normal sterile fashion. The lower chest and entire abdomen are draped into the field to allow maximal access.

Step 3. The positions for port sites are marked approximately 1-2 fingerbreadths below the costal margin extending from the posterior axillary line to the midclavicular line with at least 6 cm between the port sites. A pneumoperitoneum is then created with a Veress needle inserted through a small nick in the skin. For left adrenalectomy, the Veress needle is inserted through one of the marked port sites near the anterior axillary line. On the right side, to avoid injury to the liver, the pneumoperitoneum is created through a separate stab wound closer to the umbilicus.

Step 4. After creating the pneumoperitoneum, a 5 or 10 mm port is placed into the peritoneal cavity, depending on the size of 30° laparoscopic camera that is available. The 30° laparoscope is then inserted, and the additional three ports are placed in the positions identified. It may be necessary to take down the lateral attachments of the left colon to place the last port on the left side or mobilize a portion of the right lobe of the liver on the right side.

Northwestern Handbook of Surgical Procedures, edited by Richard H. Bell, Jr. and Dixon B. Kaufman. ©2005 Landes Bioscience.

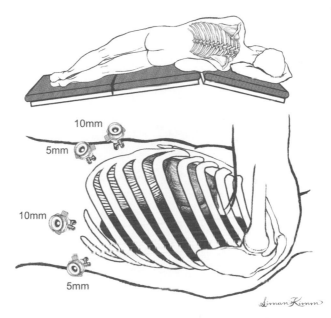

Figure 50.1. Laparoscopic adrenalectomy. Patient position and port placement.

Step 5. For left-sided adrenalectomy, the lateral attachments of the spleen are divided with a harmonic scalpel. This allows the spleen to fall medially, taking the tail of the pancreas with it and opening up the retroperitoneal space. On the right side, it is necessary to enter the retroperitoneum at the posterior aspect of the right lobe of the liver so that the liver can be retracted anteriorly. The harmonic scalpel is used to separate tissue to allow identification of the inferior vena cava.

Step 6. On the left side, the kidney is identified and the tissue superior and medial to it is inspected to allow identification of the left adrenal gland. If there is difficulty identifying the gland, a laparoscopic ultrasound probe can be used to identify an adrenal mass in the retroperitoneal fat. On the right side, the dissection involves also identifying the kidney and then identifying the adrenal gland in the tissue medial and superior to the kidney. No matter which side is being dissected, the harmonic scalpel should be used at this point to carefully dissect the tissue lateral and inferior to the adrenal gland in order to better define the extent of the gland.

Step 7. If a pheochromocytoma is present, the adrenal vein should be controlled first, by identifying the vessel, doubly clipping it, and then dividing it. The right adrenal vein is quite short and can cause significant problems with hemorrhage if not carefully dissected and divided.

Step 8. The posterior and superior attachments of the adrenal gland are divided with the harmonic scalpel, allowing the gland to be carefully separated from all of the surrounding tissues.

Step 9. Once the gland is completely separated from the surrounding tissues, it is placed within a bag inside the patient. It is then removed through one of the port sites, extending the port as necessary to allow the gland to be removed intact in the bag.

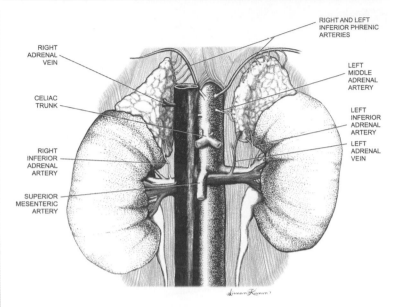

Figure 50.2. Laparoscopic adrenalectomy. Adrenal anatomy.

Step 10. The port is then reinserted into the patient for further examination of the bed of the adrenal gland. This space is inspected, irrigated, and drained of fluid to allow adequate hemostasis to be confirmed. The ports are then removed and the fascia closed on each with interrupted O Vicryl sutures. The skin is closed with monofilament absorbable subcuticular stitches.

Postop

If a pheochromocytoma has been removed, patients are observed overnight in the ICU to allow adequate fluid resuscitation as necessary and close observation of blood pressure. Most patients can be safely discharged 1-2 days after a laparoscopic adrenalectomy.

Complications

Patients should be closely followed for any signs of hemorrhage or peritonitis due to injury of any of the organs in proximity to the adrenal gland, such as the colon, spleen, or liver.

Follow-Up

Surgical sites are checked at 3 weeks postop and again at 6 months. All should be followed as appropriate to ascertain resolution of symptoms and signs (e.g., hypertension). Pheochromocytoma patients should have annual 24-hour urine sampling for VMA, catecholamines, and metanephrine levels.

Chapter 51

Pancreatic Endocrine Tumor Enucleation

Daphne W. Denham

Indications

Enucleation of a pancreatic tumor is usually performed for insulinoma. Other tumors which may be amenable to enucleation include somatostatinomas, glucagonomas, VIPomas, and nonsecretory islet cell tumors as well as serous cystadenomas. Enucleation is not appropriate for tumors with any significant likelihood of malignancy. Other factors being equal, tumors in the pancreatic head may be more attractive for enucleation than tumors in the body and tail because of the increased morbidity associated with pancreaticoduodenectomy. It is probably wise not to enucleate tumors which are intimately related to the main pancreatic duct on imaging.

Preoperative localization is ordinarily performed with some combination of endoscopic ultrasound, CT, MRI, selective venous sampling, and/or octreotide scanning, depending on the nature of the lesion. Although most insulinomas are benign, 60-90% of other islet tumors are malignant, so preoperative imaging should also document the presence or absence of metastatic disease.

Preop

In insulinoma patients, it is most important to assure that NPO status does not cause severe hypoglycemia. Intravenous fluid should be begun preoperatively and blood sugar maintained in at least the 60-80 mg/dL range.

In gastrinoma patients, active ulcers need to begin healing, with H2 blockers or proton pump inhibitors, prior to operation. A preoperative prophylactic antibiotic is given approximately 30 minutes prior to incision. Deep vein thrombosis prophylaxis with sequential compression devices or subcutaneous heparin should be employed in patients according to risk.

Procedure

Step 1. The abdomen is prepped and draped for a midline or chevron incision. In most patients, and particularly in obese patients, a chevron incision permits the best exposure.

Step 2. The abdomen is fully explored. Metastasis to the liver and regional lymph nodes must be excluded, as their presence is likely to change the planned operation. If local lymph nodes are enlarged, it is appropriate to change from an enucleation to a formal resection. Additionally, the ovaries in females must be examined for tumor implants. Although distant metastatic disease usually prohibits cure, enucleation with or without resection of metastatic deposits may be indicated for symptom control provided the patient's functional status, the extent of disease, and operative risk are taken into consideration.

Northwestern Handbook of Surgical Procedures, edited by Richard H. Bell, Jr. and Dixon B. Kaufman. ©2005 Landes Bioscience.

Step 3. The primary lesion is ordinarily identified by visualization or bimanual palpation. The exposure of the pancreas necessary for the operation may be tailored if the tumor was identified preoperatively; however, multiple tumors have been reported in sporadic cases, and it is probably advisable in most cases to carefully explore the entire gland.

Step 4. The body and tail of the pancreas is exposed by opening the lesser sac. After elevating and retracting the stomach and omentum cephalad, the omentum is taken off of the transverse colon, staying in the relatively avascular plane immediately abutting the colon. The splenic flexure may have to be mobilized to allow complete visualization of the distal portion of the pancreas.

Step 5. The body and tail of the pancreas may be additionally assessed by incising the peritoneum just below the inferior border of the pancreas and mobilizing the pancreatic tail by blunt dissection in the retropancreatic space. If necessary, the lateral attachments of the spleen may be taken down, allowing medial rotation of the spleen and tail of the pancreas and exposure of the posterior surface of the pancreas.

Step 6. To inspect the head of the pancreas, the hepatic flexure of the colon is taken down and the base of the transverse mesocolon swept inferiorly off the anterior surface of the pancreatic head. A wide Kocher maneuver is then performed to allow bimanual palpation of the head of the gland.

Step 7. Intraoperative ultrasound is very useful for visualization of the tumor's location in relation to the pancreatic duct or surrounding blood vessels. Intravenous ultrasound is also beneficial if the tumor cannot be appreciated by palpation.

Step 8. Once the tumor has been identified, using electrocautery and/or blunt dissection, the tumor is simply shelled out, staying right on the tumor capsule. If the edges of the tumor are not apparent or the tumor appears to be irregular or infiltrating, enucleation should be abandoned and a formal resection performed.

Step 9. The bed of the tumor is inspected for hemostasis and for any evidence of a major pancreatic duct injury. Any suspected ductal injury should be repaired over a stent if possible, passing the tip of the stent into the duodenum for later retrieval. If a major duct injury is present and the surgeon is unable to repair it without difficulty, it is best to proceed with resection of the involved area.

Step 10. A closed-suction drain should be placed near the enucleation site and brought out through a separate stab incision.

Postop

For insulinoma patients, glucose-free solutions should be used for intravenous fluid replacement. The blood sugar should be regularly monitored because it typically rises quickly, even while still in the operating room. Overnight, blood sugar elevations may reach the mid 200s and require a small dose of insulin. Blood sugar should be checked three times per day until stable. Patients are requested to check a fasting blood sugar daily until their follow-up clinic visit.

Patients may be fed as soon as there is return of bowel function. The drain is kept in place until the patient is tolerating food and there is no amylase-rich drainage. If there is a pancreatic leak, the drain is kept in until the fistula resolves. Somatostatin analogue injections may be helpful in reducing the quantity of pancreatic fluid from the fistula.

51

Complications

Complications of enucleation are relatively frequent and include pancreatic duct injury with pancreatic fistula and/or pseudocyst formation, peripancreatic abscess, and pancreatitis.

Follow-Up

Patients with sporadic, nonmalignant pancreatic endocrine tumors are not likely to recur. Multiple endocrine neoplasia patients often require generous distal pancreatectomy along with enucleation of tumors from the head of the pancreas and must be followed for endocrine and exocrine insufficiency. Malignant tumors require long-term follow-up for recurrent disease.

51

Chapter 52

Parathyroid Adenoma Excision

Daphne W. Denham

Indications

Excision of a parathyroid adenoma is indicated for primary and occasionally for tertiary hyperparathyroidism.

Preop

Patients should be adequately hydrated prior to induction of anesthesia. General endotracheal anesthesia is recommended.

Procedure

Step 1. The patient is placed supine, with a shoulder roll placed horizontally under both scapulae and the neck extended with the head resting on a "doughnut." The endotracheal tube should be secured away from the operative field.

Step 2. After skin preparation and draping, a transverse cervical incision is made one fingerbreadth above the clavicular heads, in a natural crease if possible. Symmetry is key to a good cosmetic result.

Step 3. Using electrocautery, the platysma muscle is divided and subplatysmal flaps raised through the superficial fascia, being careful to stay above the anterior jugular veins.

Step 4. The strap muscles are opened through the midline, typically an avascular plane. The sternohyoid and sternothyroid muscles are elevated off the anterior surface of the thyroid.

Step 5. Addressing one side at a time, the thyroid lobe is gently mobilized anteriorly and medially. Great attention to detail is necessary as a bloodless field is optimal to allow visualization of the parathyroid glands and the recurrent laryngeal nerves.

Step 6. The middle thyroid vein is identified, ligated, and divided. Additional surrounding tissues are bluntly dissected with either the surgeon's index finger or a "peanut" dissector, pushing the tissue dorsally and laterally while continuing to rotate the thyroid gland up and out of the field.

Step 7. The recurrent laryngeal nerve (RLN) is identified. The right RLN is found medial to the carotid, traveling obliquely from lateral to medial, from deep to superficial. The left RLN is typically in the tracheoesophageal groove, running in a more vertical direction.

Step 8. With the nerves identified, a systematic search for the parathyroid glands is begun. Normal parathyroid glands (PT) are typically 4-6 mm in length, 2-4 mm in width, weigh 40-60 mg, and are mustard brown in color.

Step 9. The superior PT is usually located just above the entrance of the inferior thyroid artery into the thyroid gland. It is typically *posterior* and superior to the recurrent laryngeal nerve, and most often found behind the upper two-thirds of the

Northwestern Handbook of Surgical Procedures, edited by Richard H. Bell, Jr. and Dixon B. Kaufman. ©2005 Landes Bioscience.

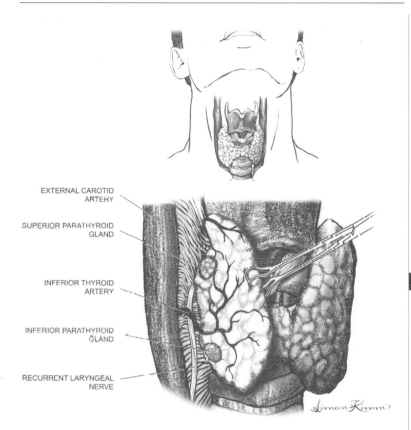

EXTERNAL CAROTID ARTERY

SUPERIOR PARATHYROID GLAND

INFERIOR THYROID ARTERY

INFERIOR PARATHYROID GLAND

RECURRENT LARYNGEAL NERVE

52

Figure 52.1. Parathyroid adenoma.

thyroid gland. Enlargement of a superior PT may cause it to drop inferior to the inferior thyroid artery, into the retropharyngeal space or into the posterior mediastinum. Typically these aberrant glands are best identified by looking for a pedicle with an obvious blood supply tracking down, as most often even the superior gland blood supply is from the inferior thyroid artery.

Step 10. The inferior PT is typically *anterior* to the recurrent laryngeal nerve, most often within 2 cm of the inferior pole of the thyroid gland. It can be in the thyrothymic ligament, which is best identified by finding the tongue of the thymus (a vascularized pedicle of fatty tissue extending in a caudal direction) and mobilizing it into the field. The inferior gland, however, may be located anywhere from the angle of the mandible to the arch of the aorta. It is the gland with the greatest potential for aberrancy.

Step 11. An enlarged PT should "roll" under the overlying connective tissue, whereas lymph nodes and thyroid nodules are typically more "fixed" to their surrounding structures. Observation of this phenomenon during blunt dissection is a key to this operation.

Step 12. Unless a preoperative localization study has been performed, an attempt should be made to identify all four glands prior to removal of any parathyroid tissue. One obvious large gland with three normal-appearing glands is consistent with a single adenoma.

Step 13. Once identified, the adenoma is best removed by gently teasing away or splitting the overlying tissue with a right angle or hemostat. The gland should essentially "pop" out. After gently grasping the distal end, trying not to rupture the capsule, a clip is applied to the pedicle and the gland is removed. Difficult dissection or a thick fibrous capsule should raise consideration of a parathyroid cancer, which requires en bloc resection.

Step 14. A meticulous search to assure hemostasis is performed. If in doubt, a small drain can be placed.

Step 15. The strap muscles are reapproximated with interrupted absorbable suture, followed by the platysma layer. A subcuticular skin closure is performed.

Postop

Most patients can be discharged within 23 hours. Diet is reinstituted as tolerated. Patients are started on oral calcium supplementation, beginning at 1 g per day, which can be increased up to 2.5 g if necessary. Patients are instructed to call immediately with symptoms of perioral or other numbness and tingling.

52

Complications

Complications of parathyroidectomy include hypocalcemia, recurrent laryngeal nerve injury, neck hematoma, wound infection, and missed adenoma.

Follow-Up

Serum calcium levels should be followed until they normalize, at which time calcium supplementation can be discontinued. Serum calcium levels are monitored yearly and bone density scanning done on a routine basis.

Chapter 53

Radioguided Parathyroidectomy: Minimally Invasive

Daphne W. Denham

Indications

Minimally invasive radioguided parathyroidectomy is indicated for primary hyperparathyroidism, when it is non-familial and non-MEN-related.

Preop

Patients should be well hydrated prior to induction of anesthesia. Local anesthesia with monitored intravenous sedation is well tolerated by most patients; however, general anesthesia is recommended for patients with a deep adenoma.

Patients must have a positive sestamibi scan which consists of an anteroposterior view including the mediastinum, a right anterior oblique view, and a left anterior oblique view. A positive scan is defined as one showing a single "hot spot" which can be distinguished from the thyroid gland in the oblique views. If the patient has a negative scan, then a standard four-gland exploration is recommended.

The patient should be in the operating room within 2 hours of injection to allow for thyroid uptake to diminish, yet parathyroid uptake to remain.

Procedure

Step 1. The patient is positioned supine, with a shoulder roll placed horizontally under both scapulae and the neck extended with the head resting on a "doughnut." If local anesthesia with sedation is being employed, 1% lidocaine with epinephrine is injected at the incision site and then into deeper tissues as the procedure progresses.

Step 2. A 2 cm midline incision is made one fingerbreadth above the clavicular heads. The incision may be placed slightly higher or lower based on the location of adenoma on the scan.

Step 3. Subplatysmal flaps are raised to the level of the cricoid and laterally to sternocleidmastoid muscles. Good flaps will allow maximal exposure through the small incision.

Step 4. The strap muscles are opened in the midline and the thyroid gland is exposed on the side of the adenoma.

Step 5. A gamma detector probe is placed through the incision to obtain counts from the thyroid (background) and in the direction of the adenoma (as suggested by the sestamibi scan). Experience is required to develop expertise with use of the probe as the adenoma is a "relative" hot spot against a "warm" background. Medial rotation of the thyroid gland is necessary to obtain counts of the tissue posterior to the thyroid.

Northwestern Handbook of Surgical Procedures, edited by Richard H. Bell, Jr. and Dixon B. Kaufman. ©2005 Landes Bioscience.

Step 6. Gentle, blunt dissection is performed in the direction guided by the probe. The oblique views of the sestamibi scan should give the surgeon an idea of the depth of the adenoma.

Step 7. After identification of the adenoma, it is gently teased from the surrounding tissues, the pedicle is clipped, and the adenoma is removed.

Step 8. The adenoma is placed over the probe, away from the operative field for an ex vivo count. If the count is greater than 20% of background, a frozen section is not necessary.

Step 9. The probe is reinserted into the wound to assure the surgeon that the "hot spot" was removed. Background counts should now be less than starting counts.

Step 10. Hemostasis is obtained.

Step 11. The strap muscles are closed with interrupted absorbable suture, the platysma (superficial fascia layer) is closed with running absorbable suture, and the skin is closed with a subcuticular suture.

Postop

Patients may be discharged from the hospital the same day, with careful instructions. Patients are begun on calcium supplementation, beginning at 1 g per day, which can be increased up to 2.5 g if necessary. Patients are instructed to call immediately with symptoms of perioral or other numbness and tingling. The incision is kept clean and dry for at least 24 hours, and then the patient is allowed to shower and pat dry the incision.

53

Complications

Complications include hypocalcemia, wound hematoma, wound infection, recurrent laryngeal nerve injury, and missed adenoma.

Follow-Up

After serum calcium levels normalize, calcium supplements can be discontinued. Serum calcium levels are followed on a yearly basis and bone density scanning done on a routine basis.

Chapter 54

Thyroid Lobectomy and Total Thyroidectomy

Peter Angelos

Indications

A thyroid lobectomy is most commonly indicated when a thyroid mass is present and cancer cannot be ruled out. In such situations, the treatment of cancer or possible cancer is the indication for surgery. Occasionally, the patient will have symptoms from a large thyroid mass. A small percentage of patients with hyperthyroidism will require surgery if medication or radioactive iodine are not options. A total thyroidectomy is indicated for the treatment of cancer, as well as for Graves' disease if the patient has not responded to medication appropriately.

Preop

It is essential preoperatively to know the patient's calcium level. In addition, if the patient has ever had previous thyroid or parathyroid surgery, it is essential to examine the vocal cords for bilateral function to rule out a unilateral recurrent laryngeal nerve injury.

Step 1. After the induction of general endotracheal anesthesia, the patient's neck is extended on a long beanbag that supports the neck in full extension. The patient is placed in semi-Fowler's position to decompress the neck veins. The bed is turned 90 degrees from the anesthesiologist to give the surgeon access all around the head. A flexible ether screen is used to protect the patient's face and allow the endotracheal tube to be secured

Step 2. The entire neck up to the chin and laterally to the shoulders and down onto the upper chest is prepped. Two crushed towels are used and then four towels extending onto the ether screen. A "U" drape is used to cover the patient.

Step 3. A low transverse collar incision is made 1-2 fingerbreadths above the clavicular heads. The incision should be centered in the midline, in a skin crease if possible. Most thyroid resections can be done safely through 5-6 cm incisions.

Step 4. After dissecting down to below the platysma, large subplatysmal flaps are created with electrocautery. It is helpful to use a needle tip electrocautery in order to allow precise tissue dissection. The limits of the subplatysmal flaps are the notch of the thyroid cartilage superiorly, the clavicular heads inferiorly, and the sternocleidomastoid muscles bilaterally.

Step 5. The midline is opened with electrocautery, allowing the strap muscles to be retracted laterally. This allows exposure of the thyroid gland. Initially dissection should be undertaken on the side with the tumor or nodule. A middle thyroid vein, if present, should be divided. This allows the thyroid gland to be rotated medially and facilitates the separation of the lateral aspect of the thyroid gland from the surrounding tissue.

Northwestern Handbook of Surgical Procedures, edited by Richard H. Bell, Jr. and Dixon B. Kaufman. ©2005 Landes Bioscience.

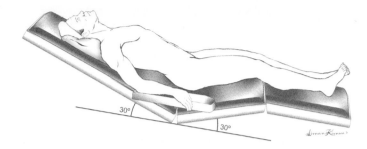

Figure 54.1. Thyroid lobectomy and total thyroidectomy. Patient position.

Step 6. Attention is initially directed toward the superior pole of the thyroid gland. The lateral edge of the thyroid gland is freed up all the way to the top of the superior pole. An avascular plane medial to the superior pole of the thyroid gland is entered to allow the superior pole of the thyroid to be retracted in a caudal direction. This allows careful exposure of the superior pole vessels as they enter the thyroid capsule. The vessels are divided and ligated individually in order to prevent injury to the external branch of the superior laryngeal nerve. A harmonic scalpel may also be safely used to cauterize and divide these and other blood vessels.

Step 7. After dividing the superior pole vessels, the thyroid lobe is pivoted medially, allowing exposure of inferior pole vessels. These vessels should be divided and ligated also as they enter the thyroid capsule.

Step 8. With the superior and inferior vessels divided, the entire thyroid lobe can be rotated medially. The parathyroid glands are then identified and carefully dissected off of the capsule of the thyroid gland. If the parathyroid glands are dissected from a medial to lateral direction, the blood supply is generally well protected.

Step 9. Once the parathyroid tissue is freed and identified, gentle blunt dissection in the tracheoesophageal groove will expose the recurrent laryngeal nerve. Once the nerve is identified, branches of the inferior thyroid artery are individually divided and ligated taking care to preserve the recurrent laryngeal nerve intact. Care should be taken to ensure that the tubercle of Zuckerkandl (that portion of the thyroid gland that extends posterior to the recurrent laryngeal nerve) is carefully elevated from its position posterior to the recurrent laryngeal nerve so that no significant thyroid tissue is left.

Step 10. Once the recurrent laryngeal nerve has been safely freed from close proximity to the thyroid capsule, the ligament of Berry is carefully divided, making sure that the dissection plane is right on the surface of the trachea. If one is performing a thyroid lobectomy and isthmusectomy, the dissection is carried over to the contralateral side where the thyroid isthmus is divided. The cut edge of the thyroid gland is oversewn with interrupted 4-0-silk figure-of-eight sutures for hemostasis.

Step 11. The bed of the resected thyroid is irrigated and carefully inspected for hemostasis. The viability of the parathyroid glands is ensured. Should a parathyroid gland appear to be completely devascularized, it should be removed from the patient, carefully minced into small pieces, and autotransplanted into a pocket into the sternocleidomastoid muscle. This pocket should be oversewn with a 4-0-silk stitch in a figure-of-eight fashion and marked with a titanium clip.

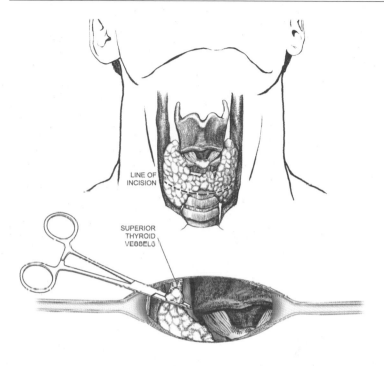

LINE OF
INCISION

SUPERIOR
THYROID
VESSELS

54

Figure 54.2. Thyroid lobectomy and total thyroidectomy.

Step 12. If a malignancy is found, lymph nodes in the central compartment on the side of the mass should be carefully palpated. Any abnormal lymph nodes should be removed. Care should be taken to try to identify the parathyroid gland in proximity to this lymph node packet prior to removing the nodes.

Step 13. If a total thyroidectomy is to be performed, the dissection is then undertaken on the contralateral side in a similar fashion to that described on the first side. The middle thyroid vein is divided, the thyroid gland is rotated medially, the superior then inferior pole vessels are individually ligated and divided at the level of the thyroid capsule. Parathyroid glands are identified and carefully dissected off of the capsule of the thyroid gland, and the thyroid tissue is elevated from deep in the tracheoesophageal groove. On this side, as described on the previous side, branches of the inferior thyroid artery are then individually ligated and divided at the level of the thyroid capsule taking care to preserve the recurrent laryngeal nerve intact. The ligament of Berry is then again meticulously dissected such that all grossly visible thyroid tissue is removed from the patient. Again, the wound is irrigated and drained of fluid.

Step 14. The deep strap muscle layer is closed with one or two interrupted simple 3-0 absorbable stitches. The superficial strap muscles are closed by reapproximating the overlying superficial fascia with interrupted 4-0-silk sutures. The platysma is then closed with interrupted 4-0-silk sutures with buried knots. The skin and subcutaneous tissue are then infiltrated with a long-acting local anesthetic and the skin is closed with a 5-0 monofilament absorbable suture in a subcuticular fashion. Steri-strips are placed lengthwise over the incision and a single Telfa dressing is placed over the steri-strips. If adequate adhesive is placed on the skin surrounding the incision, this small dressing will be held in place without the need for any tape. This small neck dressing allows rapid identification of hematoma should one develop. This outer dressing can also be readily removed the morning after surgery and the patient is then discharged with steri-strips in place.

Addsteps. If a very large thyroid gland has been removed or if there has been an extensive lymph node dissection, the surgeon may opt to place a 4 or 7 mm closed suction drain taken out through a separate stab wound. For cosmetic reasons, it is best for the drain to exit the skin in the midline through a small transverse incision directly below the operative incision. The drain can almost always be removed on the day after surgery.

Postop

After thyroidectomy, we routinely admit patients for an overnight stay. If respiratory distress with stridor develops, the neck incision should be opened at the bedside with a presumptive diagnosis of hematoma.If the patient has undergone a total thyroidectomy, we obtain an ionized calcium level at 16 hours after the completion of the operation. Based on this level, a decision on whether calcium supplementation is necessary is made.

Complications

The primary complications of thyroidectomy are neck hematoma, temporary or permanent hypocalcemia, and unilateral or bilateral recurrent or superior laryngeal nerve injury.

Follow-Up

Patients are seen for an initial follow-up visit at approximately 3 weeks postoperatively and then again at 6 months if there were no complications. If a thyroid lobectomy was performed, approximately 50% of patients will require thyroid hormone replacement, which can be determined by monitoring the TSH level 4-6 weeks after surgery.

54

Chapter 55

Modified Neck Dissection

Peter Angelos and Jeffrey D. Wayne

Indications

The extent of lymph node dissection recommended for thyroid cancer is a controversial topic. Specifically, questions surround the optimal extent of lymph node dissection for well-differentiated thyroid cancer. There is no controversy about the need for an extensive lymph node dissection when dealing with medullary thyroid cancer. However, only approximately 10% of thyroid cancers are of the medullary type. For the more common and less biologically aggressive papillary and follicular thyroid cancers, the issue is how much of a neck dissection is enough. The choices in lymph node dissection vary from a minimalist node sampling operation to a classic radical neck dissection. Most authors agree that there is no need for a radical neck dissection because equally good local control and cure are readily obtained with less extensive operations. The modified radical neck dissection (or functional neck dissection) allows en bloc dissection of the lymphatic system in levels I through IV while preserving the sternocleidomastoid muscle, jugular veins, and/or spinal accessory nerve.

Neck dissection is indicated in patients with papillary and follicular thyroid cancers that have enlarged cervical lymph nodes where metastases are expected. Neck dissection is required for all patients with medullary thyroid carcinoma. The following operative approach may also be applied to patients with melanoma who have either clinically involved cervical lymph nodes upon presentation or, more commonly, who have a positive sentinel lymph node biopsy.

Preop

With thyroid cancer, the lymph node dissection is commonly performed at the time of total thyroidectomy. It is always important to check the status of the patient's vocal cords if there has been previous neck surgery or if the patient has preoperative hoarseness.

Procedure

Step 1. The operating room setup is the same as described for a thyroidectomy. Anesthesia is general by endotracheal intubation.

Step 2. The same low transverse collar incision is used as for a thyroidectomy. If necessary, this incision is extended in a skin crease laterally on either side. If it is necessary to gain access more cranially, the incision is extended up the anterior border of the sternocleidomastoid muscle. For melanoma, a number of incisions have been described. The most common is vertical along the anterior border of the sternocleidomastoid muscle, with a short vertical limb from the midportion of the vertical incision to the level of the hyoid bone.

Northwestern Handbook of Surgical Procedures, edited by Richard H. Bell, Jr. and Dixon B. Kaufman. ©2005 Landes Bioscience.

Step 3. Adequate exposure is critical and, therefore, large subplatysmal flaps should be raised utilizing electrocautery for dissection. We recommend the use of a needle tip for greater precision.

Step 4. On the ipsilateral side of the tumor or bilaterally for medullary thyroid cancer, attention is initially directed to exposure of the insertion of the sternocleido-mastoid muscle to the sternum and clavicle. It is possible to retract the sternocleido-mastoid muscle without dividing it. However, in order to allow maximal exposure, the tendinous and muscular insertions are divided and the muscle is reflected superiorly.

Step 5. The tissue plane immediately posterior to the sternocleidomastoid muscle is divided in order to allow adequate cranial dissection. The lymph node dissection begins at the superiormost aspect near the angle of the mandible. Here, one should identify the marginal mandibular branch of the facial nerve and retract it superiorly so as to avoid injury. At this point node-bearing soft tissue should be identified and the superior extent of the dissection defined. All lymph nodes and associated adventitia above the vascular sheath are then swept inferiorly. This dissection is best performed with a combination of blunt and sharp dissection.

Step 6. As the dissection is carried in a caudal direction, the omohyoid muscle will be identified crossing the field. The omohyoid muscle should be divided at this point.

Step 7. As the dissection extends further in a caudal direction, it is important to remove all of the soft tissue anterior and adjacent to the carotid artery, the internal jugular vein, and the vagus nerve. In this fashion, lymph-node-bearing tissue from both the anterior triangle and posterior triangle of the neck can be removed. Care must be taken to avoid injury to the spinal accessory nerve in the posterior triangle.

Step 8. As the dissection approaches the clavicle, it is important to identify the thyrothymic tract. While taking care to protect the recurrent laryngeal nerve as well as the vascular supply of the inferior parathyroid gland, soft tissue of the thyrothymic tract along with the associated lymph nodes should be removed. During the course of this portion of the dissection, it is important to remove lymph nodes anterior and immediately lateral to the trachea. These paratracheal and pretracheal lymph nodes are frequently involved in thyroid cancer and should be included in the neck dissection. Also at this portion of the dissection, care must be directed to protecting the right lymphatic duct and the thoracic duct on the left side. The thoracic duct extends above the clavicle and enters the internal jugular vein near the junction with the subclavian vein. If an injury to the thoracic duct occurs, it should be identified and ligated.

Step 9. If the dissection is for medullary thyroid cancer, a bilateral dissection should be performed such that all central compartment nodal tissue is removed. The neck is then irrigated and a closed suction drain placed through a separate stab wound.

Step 10. The sternocleidomastoid muscle is then sutured to the sternal and clavicular attachments. The strap muscles, which were split during the course of the thyroidectomy, are reapproximated and the platysma is reapproximated with interrupted sutures.

Step 11. The skin is closed with a subcuticular stitch.

55

Postop

The closed suction drain should be left until there is less than 30 cc in 24 hours output. Patients may be discharged with the drain in place.

Complications

Potential complications of lymph node dissections include all of the complications associated with a thyroidectomy. These include devascularization of the parathyroid glands, recurrent laryngeal nerve injury, and the potential for hematoma. In addition, there is a small risk of lymphocele. This is unlikely if a closed suction drain is left for an adequate period of time and if care is taken not to injure the thoracic duct or right lymphatic duct without ligation. Patients can generally be discharged on the day following surgery.

Follow-Up

The drain is generally left in place until there is less than 30 cc of output in 24 hours. Check the surgical site in 3 weeks and then again in 3 to 6 months. Additional treatment of thyroid cancer is dependent upon tumor type. In the case of stage III melanoma, patients should be referred to a medical oncologist for consideration of adjuvant therapy with interferon α-2b.

Section 3: Surgical Oncology
Section Editor: Mark S. Talamonti

Chapter 56

Transanal Excision of Rectal Tumor

Amy L. Halverson

Indications

Transanal excision is appropriate for benign lesions that are not amenable to endoscopic resection and for early-stage malignant tumors in select individuals. A tumor may be considered for transanal resection if it is less than 9 cm from the anal verge, less than 4 cm in length, mobile (there should be no suggestion of anal sphincter involvement), and involves less than one-third the circumference of the rectal wall. Malignant tumors should be well or moderately differentiated and have no lymph or vascular invasion. Extension beyond the rectal wall or lymph node involvement should be ruled out with preoperative endorectal ultrasound. Transanal excision may be used for palliation in patients with overt metastatic disease.

Preop

- Complete history, including family history
- Physical examination
- Endorectal ultrasound to assess depth of invasion and lymph node involvement
- Colonoscopy to evaluate the entire colon
- CT scan to rule out metastatic disease
- Complete bowel preparation
- The patient should be placed in the prone position if the lesion is anterior. The lithotomy position may be used for posterior lesions, although some surgeons prefer the prone position for posterior lesions as well.

Procedure

Step 1. Visualize the mass through an operating anoscope.

Step 2. Score the line of resection around the mass with electrocautery; a 1 cm margin is preferable.

Step 3. Excise the lesion. A clamp may be placed on the mass for retraction. For tumors that are malignant or suspicious for malignancy, full-thickness excision should be performed. The yellow perirectal fat will be visible with full-thickness resection. For benign lesions, the submucosa may be infiltrated with saline to aid in resection of the polyp leaving the muscular layer of the bowel wall intact.

Step 4. Wounds limited to within 3-4 cm of the dentate line may be closed or left open. Higher lesions should be closed in a transverse fashion with full-thickness absorbable suture.

Step 5. Orient the specimen for the pathologist. This may be done by securing the polyp to a piece of cardboard or using different sutures to mark proximal or distal and right or left.

Step 6. After completion, double check for hemostasis.

Northwestern Handbook of Surgical Procedures, edited by Richard H. Bell, Jr. and Dixon B. Kaufman. ©2005 Landes Bioscience.

Postop

Patients should take stool softeners for approximately one week, or more if they tend to have constipation. Bleeding, urinary retention, or recurrence occur in 20% of patients.

Follow-Up

For benign or malignant lesions, repeat proctoscopy in 4-6 weeks to detect residual disease or early recurrence. For malignant tumors, surveillance includes physical examination including proctoscopy every 3 months for 2 years.

56

Chapter 57

Abdominoperineal Resection

Steven J. Stryker

Indications

Indications for abdominoperineal resection (APR) of the rectum can be categorized into absolute and relative. Absolute indications include malignancy of the rectum with sphincter involvement, carcinoma of the anal canal in an individual with prior pelvic radiation for an unrelated malignancy, carcinoma of the anal canal that is persistent or has recurred after combined modality chemotherapy and radiation, and anorectal Crohn's disease with uncontrolled local septic complications. Relative indications for APR include malignancy of the rectum not involving the sphincter when continence is already impaired preoperatively, ulcerative colitis or Crohn's proctitis requiring surgical intervention in an individual not desiring a sphincter-preserving procedure, and radiation-induced proctitis not responding to nonoperative measures or fecal diversion alone.

Preop

Preoperative preparation consists of a thorough diagnostic workup to assess the extent and severity of the disease process. This may include, but is not limited to, colonoscopy, endorectal ultrasound, and computerized tomography. If studies demonstrate unilateral or bilateral hydronephrosis, cystoscopy with placement of ureteral catheters should be considered at the time of the APR. The patient should meet with an enterostomal therapist prior to surgery, both to mark the best-suited site for colostomy placement and to provide educational materials relative to stomal function and care. Perioperative intravenous antibiotics are administered to decrease the incidence of postoperative infectious complications. The advantages of oral antibiotics on the day prior to surgery, as well as a mechanical bowel lavage, are not as well documented to decrease infection, but are nonetheless usually used in conjunction with intravenous antibiotics. All patients undergoing APR should have a type and crossmatch because of the risk, albeit low, of hemorrhage if the presacral venous plexus is disrupted intraoperatively. Thigh-high graded compression stockings along with pneumatic compression sleeves for the lower extremities are used to minimize the risk of deep venous thrombosis. In patients with a prior history of venous thrombosis or pulmonary embolus, subcutaneous heparin may be used as well.

Procedure

Step 1. The patient is positioned in a dorsal lithotomy fashion allowing simultaneous access to the abdomen and perineum. It is preferable to position the patient awake to check for comfort in positioning with respect to the back, hips, and knees.

Northwestern Handbook of Surgical Procedures, edited by Richard H. Bell, Jr. and Dixon B. Kaufman. ©2005 Landes Bioscience.

Once the patient confirms that the positioning is comfortable, general anesthesia is induced. Care should be taken that there is no excessive pressure on the calves or the lateral aspect of the proximal leg after positioning to avoid compartmental syndrome or peroneal nerve injury postoperatively.

Step 2. The abdomen and perineum are widely prepped with a povidone/alcohol combination prep. The preoperatively chosen stoma site should be scratched with an 18-gauge needle to facilitate intraoperative localization. The rectum is irrigated with a dilute povidone solution and then the anal verge is sewn shut with a heavy silk suture in a pursestring fashion.

Step 3. The abdomen and perineum are draped to provide wide access to these areas using the particular drape combinations available to the surgeon.

Step 4. A lower midline incision is made taking care to divide the midline fascia down to the pubic symphysis. A thorough abdominal exploration is undertaken to assess the extent of tumor involvement.

Step 5. The lateral and medial peritoneal reflections of the sigmoid colon are incised down to and across the rectovesical or rectovaginal reflection. The left ureter is identified and displaced laterally along with the gonadal vessels.

Step 6. The superior hemorrhoidal vessels and distal sigmoid vessels are ligated proximally, taking care to identify and avoid the left ureter throughout this maneuver.

Step 7. The rectum and mesorectum are sharply mobilized en bloc off the sacrum and lateral pelvic sidewalls, staying on the visceral aspect of the endopelvic fascia. Waldeyer's fascia is divided posteriorly. In a male, Denonvillier's fascia is divided anteriorly as the rectum is separated off the seminal vesicles and prostate. In a female, the rectum is mobilized off the posterior vaginal wall. These planes of dissection are separated down to the levator musculature circumferentially. At this point, a cloth pack is placed deep in the pelvis between the rectum and coccyx.

Step 8. Following complete mobilization of the rectum, the proximal margin of transection is chosen, typically in the proximal one third of the sigmoid. The colon is divided at this point with a linear stapling device.

Step 9. The operating surgeon relocates to the perineum at this point and makes a circumanal incision. The incision is deepened, extending into the ischiorectal fossae bilaterally. The anococcygeal raphe is divided in the posterior midline as the posterior three quarters of the anus is mobilized in a cephalad direction using electrocautery. The inferior hemorrhoidal vessels are encountered in the anterolateral region of the ischiorectal fossae at the level of the upper anal canal. These vessels usually require separate ligation.

Step 10. When the levator ani muscles are reached from the perineal dissection, they are incised in the posterior midline, using the previously placed pack as a guide. After entering the pelvis from below, the levators are incised bilaterally from posterior to anterior until a large defect exists in the pelvic floor.

Step 11. The proximal end of the specimen is grasped from below and delivered through the perineal wound. The anterior portion of the perineal dissection is now completed by carefully incising the perineal body and continuing this dissection cephalad to the levators. In a male, this requires division of the rectourethralis ligament. In a female, the transverse perineal musculature is incised. When the pelvis is reached, the dissection is complete and the specimen is sent to pathology.

Step 12. The perineal wound is carefully inspected and hemostasis achieved. The wound is closed in layers, separately reapproximating the levators (if possible), the ischiorectal fat, and the perineal skin.

Step 13. The pelvis is inspected once again from above and irrigated. A suction drain is placed in the presacral space and brought out through the anterior wall. No attempt is made to reapproximate the residual pelvic peritoneum.

Step 14. The previously marked colostomy site is excised and the proximal sigmoid end exteriorized through this transrectus opening. The colonic segment is sutured to the anterior abdominal wall from within the peritoneal cavity.

Step 15. The midline wound is closed and the colostomy is matured by excising the staple line and sewing the full thickness of the bowel wall to the dermis circumferentially.

Postop

Intravenous antibiotics are continued for 24 hours postoperatively. The patient ambulates on the day following surgery. A urinary catheter is left in for 3-5 days. Clear liquids are begun orally upon resumption of bowel activity.

Complications

The most common early postoperative complications encountered include atelectasis, urinary tract infection, abdominal or perineal wound infection, or prolonged ileus. Late complications include peristomal herniation and adhesive small bowel obstruction.

Follow-Up

Cancer patients are seen at 3-month intervals for the first 2 years, at 4-month intervals for the 3rd year, and at 6-month intervals subsequently.

Chapter 58

Right Hepatic Lobectomy

Alan J. Koffron

Indications

Extirpation of primary hepatic malignancies (cure), or removal of metastatic lesions of the liver (prolong survival) including hepatic segments 5-8. Consider the general condition of the patient and, in particular, the functional liver status and the adaptation of the volume of resection to the size of the liver tumor.

Lobectomy is indicated in *non-cirrhotic* patients in whom clearance of tumor or benign lesions can be obtained without compromising hepatic arterial and portal venous inflow or biliary drainage and in *cirrhotic* patients with adequate liver function (Child's class A, serum bilirubin <2 mg/dl, no ascites).

Preop

- Preoperative imaging to determine the nature of the disease, native hepatic anatomy, and index of resectability.
- Preoperative transcutaneous biopsy of liver masses is not always advisable given 1) the clinical presumptive diagnosis, 2) availability of serum tumor markers, 3) accurate diagnostic imaging, and 4) risk of tumor dissemination.
- Correction of anemia and of coagulopathy and appropriate single dose of antibiotic prophylaxis (e.g., cephazolin). Patients with a history of cardiorespiratory disease and all patients over 65 years are submitted to full investigation.
- Type and crossmatch for two units of packed red blood cells. Encourage patients, particularly those with metastatic colorectal cancer to the liver, to donate two units of autologous blood prior to surgery.
- Use appropriate deep venous thrombosis prophylaxis.
- Once general anesthesia is accomplished (supine position), place nasogastric tube, Foley catheter, and invasive monitoring (arterial or central venous line) if indicated. Prep skin from nipple level to pubis.

Procedure

Step 1. Incision: long right subcostal extending across the left rectus muscle with a superior midline extension. Use a retractor system that provides elevation of the right costal margin to aid in right lobe mobilization and visualization of both the suprahepatic inferior vena cava (IVC) and retrohepatic IVC.

Step 2. Ligate and divide the ligamentum teres hepatis. Using diathermy, divide the falciform ligament up to the coronary ligament superiorly.

Step 3. Explore the abdomen thoroughly to exclude extrahepatic malignancy.

Step 4. The line of resection of liver extends from the gallbladder fossa anteriorly to the vena cava posteriorly (Cantlie's line). Ensure this line permits adequate resection margin (using bimanual palpation or intraoperative ultrasound).

Northwestern Handbook of Surgical Procedures, edited by Richard H. Bell, Jr. and Dixon B. Kaufman. ©2005 Landes Bioscience.

Step 5. The gallbladder is removed after division of the cystic artery and duct (unless adjacent to tumor, where it should be rotated laterally and kept en bloc with the right lobe specimen).

Step 6. At the base of the gallbladder fossa, the right primary branches of the hepatic artery, portal vein, and bile ducts* must be exposed and identified with great care to avoid injury to the structures serving the remnant liver (left lobe). Ligation of the hepatic artery and right hepatic duct is accomplished, and the right portal vein is best stapled (endo-GIA stapler with 2.8 mm vascular load) to prevent suture dislodgment or coarctation of the portal vein bifurcation. A line of color demarcation will be seen on the liver corresponding to the anatomic division between the right and left lobes (main portal scissura or Cantlie's line).

Step 7. Using diathermy, divide the remaining coronary ligament to expose the right hepatic vein as it enters the IVC at the base of segment 8 superiorly.

Step 8. Mobilize the right lobe by dividing the right triangular ligament.

Step 9. Continue by elevating the lobe medially, mobilizing the right lobe at the bare area (adherent to the diaphragm posteriorly). Use caution:

1. superiorly as the right hepatic vein is close to the capsule/diaphragmatic junction
2. inferiorly as the adrenal gland will be elevated with the lobe until lowered
3. medially where the IVC meets the posterior aspect of the liver

Step 10. Ligate and divide the short hepatic veins and caudate veins directly entering the retrohepatic IVC until the right lobe specimen is free medially. Divide the inferior vena caval ligament to expose the inferior margin of the right hepatic vein and aid in cava-hepatic separation.

Step 11. Replace the right lobe and begin liver parenchymal transection in an anteroinferior to posterosuperior fashion. There are many different techniques to divide the parenchyma including:

1. finger fracture with suture hemostasis
2. sequential parenchymal ligation/division
3. harmonic scalpel
4. CUSA (Cavitron Ultrasonic Aspirator) in combination with Argon beam co-agulation
5. sequential parenchymal stapling
6. inflow occlusion (Pringle maneuver) in combination with above options

Option 4 is the preference of the authors and many hepatobiliary surgeons in that the transection is performed accurately, vasculobiliary structures individually ligated, and blood loss is reduced. The CUSA, moved back and forth as a pen, removes parenchyma sparing stroma and vasculobiliary structures which are ligated; the raw surface is then coaglated with the Argon beam coagulator.

NOTE: As the transection advances deeper into the interlobar plane, be aware that the right anterior sector vein (draining segments 5 and 8 into the middle hepatic vein) crosses the transection plane and must be identified, dissected, and ligated definitively. This vein is large (5-10 mm) and is 2-3 cm deep, coursing obliquely into the inferior aspect of segment 4a.

Step 12. After completing transection, control and divide the right hepatic vein either through clamp/suture or stapling, being careful to avoid injury to the IVC or middle hepatic vein.

---

*Many surgeons prefer to divide the right hepatic duct in the liver parenchyma (during transection) to avoid hepatic duct confluence complications.

Step 13. Where appropriate, send the right lobe specimen to surgical pathology to assess resection margin.

Step 14. Survey liver cut surface for both hemostasis and biliary leak. Control with sutures. unless adjacent to middle hepatic vein and porta hepatis where caution is necessary.

Step 15. Irrigate cavity with saline, and inspect cut surface once again. Biliary leaks (4% of liver resections of any magnitude) are subtle and difficult to identify.

Step 16. Closed suction drainage of the cavity is controversial. It is unnecessary in most cases (no reduction in complications) unless there is a leak that is difficult to control or in cases where a biliary reconstruction is necessary.

Step 17. Examine liver remnant for viability and risk of torsion into the right hepatic fossa. If torsion is possible, reconstruct falciform ligament to support left lobe using absorbable suture material.

Step 18. Close incision in layers with suture material of choice with particular attention to approximation of tissues at apex (midline, center).

Postop

Monitor for hemorrhage and hepatic decompensation in early postoperative period. Remove nasogastric tube and Foley catheter on day 1. Institute diet as tolerated. Continue deep venous thrombosis and pulmonary prophylaxis. Remove drains after the patient is taking diet if non-bilious effluent.

Complications

Biliary leakage (4% of patients). Continue drainage (90% will resolve) unless high output where ERCP-stent should be considered. Fever and leukocytosis warrant imaging to exclude biloma (requires drainage).

Follow-Up

The patient should be followed until fully recovered or, if applicable, bile leak subsides. Consider contacting medical oncologist for follow-up where appropriate.

58

Chapter 59

Axillary Lymphadenectomy

Kevin Bethke

Indications

Axillary lymphadenectomy is performed for the treatment of metastatic cancer (usually breast or melanoma) to the axilla as manifest by clinical adenopathy or a positive sentinel lymph node biopsy.

Preop

Preoperative lymphoscintigraphy may be necessary if dissection is performed for melanoma of the trunk and the nodal drainage pattern is ambiguous.

Procedure

Step 1. A soft roll is placed longitudinally beneath the shoulder and flank to displace the axilla anteriorly and allow easier access to the axilla. The arm is placed at slightly less than a 90° angle from the trunk. Care is taken to prevent hyperextension of the shoulder by elevating the arm on pillows and then securing it to an arm board in a stationary position.

Step 2. A 4-5 cm, gently curved incision is made at the inferior axillary hairline between the lateral border of the pectoralis major muscle and the anterior border of the latissimus dorsi muscle.

Step 3. The lateral border of the pectoralis major and anterior border of the latissimus dorsi muscles are exposed. Dissection proceeds along the latissimus dorsi muscle superiorly to the point where the axillary vein crosses the latissimus tendon. Care is taken to protect the intercostobrachial nerve which runs transversely through the mid-axilla.

Step 4. The axillary fascia is incised inferior to the axillary vein. Gentle blunt dissection using a Kittner dissector is utilized along with constant inferior retraction to dissect the fatty, node-bearing tissue off the axillary vein. The axillary vein should not be completely stripped of all tissue because this would disrupt lymphatic vessels which course along the vein and drain the arm, leading to an increased risk of lymphedema.

Step 5. The axillary tissue is bluntly dissected away from the serratus anterior muscle medially and the lattissimus dorsi muscle laterally. The long thoracic and thoracodorsal nerves are identified and protected. The medial pectoral nerve is identified along the superior lateral border of the pectoralis major muscle and protected.

Step 6. Once the long thoracic and thoracodorsal nerves have been identified, the level II nodes posterior to the pectoralis minor muscle are retracted inferiorly and the axillary contents are swept inferiorly along the groove between the serratus anterior and lattissimus dorsi muscles. If the level III nodes medial to the pectoralis minor muscle are grossly involved they are also dissected and swept inferiorly.

Northwestern Handbook of Surgical Procedures, edited by Richard H. Bell, Jr. and Dixon B. Kaufman. ©2005 Landes Bioscience.

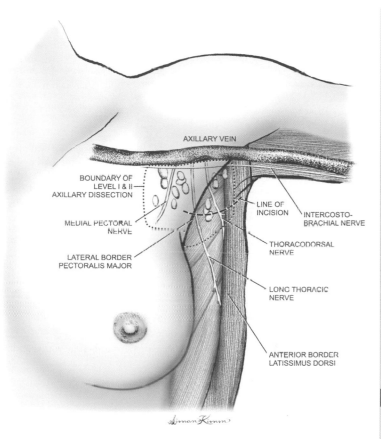

AXILLARY VEIN

BOUNDARY OF
LEVEL I & II
AXILLARY DISSECTION

MEDIAL PECTORAL
NERVE

LATERAL BORDER
PECTORALIS MAJOR

LINE OF
INCISION

INTERCOSTO-
BRACHIAL NERVE

THORACODORSAL
NERVE

LONG THORACIC
NERVE

ANTERIOR BORDER
LATISSIMUS DORSI

59

Figure 59.1. Axillary lymphadenectomy.

Step 7. The intercostobrachial sensory nerve, if not grossly involved with tumor, is dissected free of nodal tissue and retracted anteriorly. The axillary contents are then swept inferiorly and behind the nerve. Care is taken to avoid injury to the long thoracic nerve in the inferior axilla where the nerve enters the serratus anterior muscle. The larger veins draining the inferior axilla are ligated.

Step 8. A closed suction drain is placed in the axilla and brought out via a stab wound placed inferior to the incision and sutured in place.

Step 9. Bupivacaine is injected along the incision for prolonged postoperative pain relief and the wound is closed in two layers utilizing interrupted absorbable suture for the subcutaneous layers and a running 4-0 subcuticular stitch with steri-strips for the skin.

Postop

The patient is usually discharged 2-3 hours after the operation with a prescription for oral narcotic analgesia. Drain care instructions are given prior to discharge and include stripping the drain 2-3 times per day to help prevent clots from obstructing it. The patient will empty the suction bulb 2-3 times per day and record the output. When the output falls to 30-40 cc per 24 hours the drain is removed in the clinic. The patient may shower on postoperative day one and begin gentle arm and shoulder exercises.

Complications

Upper medial arm numbness can be avoided by careful dissection and preservation of the intercostobrachial nerve. Axillary seromas can be minimized by leaving the drain in until the output is down to 30-40 cc per 24 hours (usually 7-10 days). Injury to the long thoracic nerve causes dysfunction of the serratus anterior muscle and a "winged scapula." This is a very rare complication and can be minimized by adequate exposure and gentle blunt dissection of the axillary contents off the serratus muscle.

Clinical lymphedema occurs in 10-15% of patients. The more extensive the dissection performed, the greater the risk. Denuding the axillary vein during the superior axillary dissection will increase this risk, as will removal of the level III nodes. Patients should be referred to an experienced physical therapy department at the earliest sign of lymphedema.

Follow-Up

The patient should be followed by the surgeon until the wound is healed and the drain is removed. Some form of adjuvant therapy (chemotherapy and/or radiation therapy) will likely be required after which the patient will need to be followed indefinitely for signs of recurrence.

59

Chapter 60

Inguinal Lymphadenectomy

Kevin Bethke

Indications

Inguinal lymphadenectomy is performed for the treatment of metastatic melanoma (and other cancers) to the groin as manifest by clinically involved lymph nodes without other evidence of distant metastasis or by a positive sentinel lymph node biopsy.

Preop

Preoperative lymphoscintigraphy is indicated if the primary cancer is located on the trunk and the nodal drainage pattern is ambiguous. All patients are at high risk for postoperative lower extremity lymphedema and should be measured preoperatively for thigh-high compression hose. Intravenous cefazolin is given 30 minutes prior to incision. Deep vein thrombosis prophylaxis with compression boots or heparin is provided according to the degree of patient's risk.

Procedure

Step 1. The patient is placed in a supine position and general endotracheal anesthesia is induced. A pillow is placed beneath the knee, and the leg is slightly externally rotated.

Step 2. After prepping and draping, a gently curved incision located parallel to and below the inguinal crease is made, extending from the anterior iliac spine to several centimeters below the pubic tubercle.

The extent of groin dissection required can be superficial or deep. Steps 3-6 describe a superficial groin dissection; step 7 describes a deep dissection.

Step 3. The superior skin flap is raised and the nodal tissue overlying the external oblique aponeurosis and inferior to the inguinal ligament is resected. Care must be taken to resect all nodal tissue, especially if the primary site is on the trunk, superior to the incision.

Step 4. With medial traction, dissection begins at the lateral border of the sartorius muscle and proceeds towards the femoral sheath. The femoral nerve and lateral femoral cutaneous nerve lie deep to the sartorius fascia and are protected.

Step 5. The dissection proceeds over the femoral vessels in a subadventitial plane. The saphenous vein is ligated at the saphenofemoral junction. With lateral traction, the adductor longus fascia is incised and dissection proceeds medially towards the femoral sheath.

Step 6. The saphenous vein is ligated inferiorly at the apex of the femoral triangle. The only remaining attachment is at the femoral ring where Cloquet's node resides, medial to the femoral vein. The fascia overlying the femoral ring is incised and Cloquet's node is dissected free from the properitoneal fat.

Northwestern Handbook of Surgical Procedures, edited by Richard H. Bell, Jr. and Dixon B. Kaufman. ©2005 Landes Bioscience.

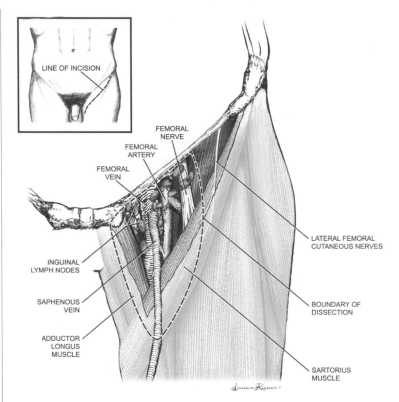

Figure 60.1. Inguinal lymphadenectomy.

60

Step 7. (Deep groin dissection). In some situations a more radical (ilioinguinal) groin dissection is indicated based on the status of Cloquet's node or significant inguinal adenopathy. Exposure for the deep dissection can be gained either by transecting the inguinal ligament as it crosses the femoral canal or by retracting the superior skin flap and making an incision through the abdominal wall superior and parallel to the inguinal ligament. The peritoneum is retracted superiorly and medially and the retroperitoneal node-bearing tissue is resected from the femoral ring (inferiorly), bladder (medially), genitofemoral nerve (laterally), and aortic bifurcation (superiorly). The inguinal ligament is then reapproximated (if transected) and/or the abdominal wall muscles are anatomically approximated. A drain is not necessary for the deep dissection.

Step 8. At the conclusion of dissection, the sartorius muscle is detached from the iliac spine and transposed over the femoral vessels. It is sutured to the inguinal ligament.

Step 9. A closed suction drain is placed in the groin wound and brought out from the abdominal wall superior to the incision in order to avoid compromised groin tissues.

Step 10. One centimeter of traumatized skin is excised from both wound edges and the wound is meticulously closed in two or three layers.

Postop

The leg is wrapped with an elastic bandage from the foot to the thigh. The leg is elevated on pillows during the hospital stay. Ambulation is encouraged on the first postoperative day. Several days after surgery the previously custom-fitted compression stocking is placed. The groin drain is left in place until the output is 30-40 cc over a 24-hour period (usually 2-3 weeks) and then removed in the clinic

Complications

Wound problems are common, with skin ischemia, subsequent necrosis and wound infection a feared complication. The most severe complication is a femoral artery blowout in an infected groin, which fortunately is quite rare when the sartorius muscle is transposed over the vessels. Lymphedema is a common problem, especially with a combined superficial and deep dissection. Patients must be warned of this and encouraged to wear their support hose for at least 4-6 months after surgery.

Follow-Up

The patient should be followed by the surgeon until wounds are healed and any lymphedema problems are addressed. Either the surgeon or oncologist must follow the patient indefinitely for signs of recurrent tumor.

60

Chapter 61

Breast Biopsy after Needle Localization

Monica Morrow

Indications

Needle-localized breast biopsy is indicated for diagnosis of nonpalpable breast abnormalities considered suspicious for carcinoma (BIRADS 4 or 5) which are not suitable for image-guided breast biopsy. It is also indicated for removal of lesions diagnosed as atypical hyperplasia or carcinoma by core biopsy.

Preop

Immediately preoperatively, a localization wire is placed into or within 1 cm of the suspicious lesion using mammographic or ultrasound guidance. The wire insertion is done in the radiology suite using local anesthesia. The biopsy is then done in the operating room using local anesthesia with intravenous sedation.

Procedure

Step 1. Localization films (two views) are reviewed to determine the relationship of the wire to the lesion and the distance from skin entry to lesion. Proper incision placement is critical to success. The incision is placed over the abnormality, *not* at the point the localization wire enters the skin.

Step 2. The patient is placed in the supine position. The skin incision is made in Langer's lines.

Step 3. Subcutaneous fat and superficial breast tissue are divided. The wire is identified within the breast parenchyma and pulled inward through the skin into the operative field.

Step 4. The breast tissue surrounding the wire is divided with scissors until the area of the lesion is approached. Identification is facilitated by using a localization wire with a thickened distal segment and placing this segment through the lesion.

Step 5. At the level of the lesion, excision is carried out to remove the wire and the lesion, surrounded by a margin of approximately 5 mm of normal breast tissue on all sides.

Step 6. The specimen is marked with a short suture in the superior margin and a long suture in the lateral margin and sent for specimen radiography to confirm removal of the target.

Step 7. After obtaining hemostasis, the cavity resulting from the excision is irrigated with saline. The four walls of the cavity are marked with clips. No attempt should be made to reapproximate the deep breast tissue.

Step 8. The deep dermis is closed with interrupted 3-0 Vicryl sutures, knots inverted. The skin is closed with a running subcuticular 4-0 Vicryl suture, steristrips, and a light gauze dressing.

Northwestern Handbook of Surgical Procedures, edited by Richard H. Bell, Jr. and Dixon B. Kaufman. ©2005 Landes Bioscience.

Postop
This is an outpatient procedure. The dressing can be removed and the patient may shower in 24 hours.

Complications
Failure to remove target abnormality in 1-2%, hematoma, and infection.

Follow-Up
The patient should be examined in 1-2 weeks to ensure adequate healing. Other follow-up procedures are specific to pathology on biopsy.

Chapter 62

Lymphatic Mapping and Sentinel Node Biopsy

Seema A. Khan

Indications

Lymphatic mapping and sentinel node biopsy are indicated for axillary nodal staging in patients with invasive breast carcinoma. It may also be indicated in patients with duct carcinoma in situ (DCIS) diagnosed by core needle biopsy when the likelihood of microscopic invasion is high (e.g., palpable DCIS and large DCIS lesions).

Preop

Lymphatic mapping can be accomplished with either a radioactive tracer (technetium sulfur colloid), or a dye tracer (lymphazurin or methylene blue), or both. The radioactive tracer is usually injected in the nuclear medicine department one or more hours prior to surgery; the dye tracer is injected in the operating room by the surgeon. Radioactive tracer (RAT) injection can be peritumoral or intradermal, or a combination of these. Dye tracer is not injected intradermally to avoid tattooing of the skin. If image-guided wire localization is needed for nonpalpable lesions, this is performed prior to the RAT injection. The sentinel node biopsy can be performed under sedation, with local anesthesia, but general anesthesia is often preferred.

Procedure

Step 1. The patient is placed in the supine position with the ipsilateral arm extended on an arm board at a nearly right angle to the operating table. Dye tracer can be injected prior to skin preparation, or following preparation and draping. Skin preparation includes the entire breast if wide excision of the breast carcinoma is being performed at the same time. The ipsilateral arm is prepared to the elbow or lower and draped into the field.

Step 2. Five ml of dye tracer is injected in a peritumoral or retroareolar fashion; the injection site is then vigorously massaged for 5-7 minutes. If RAT mapping is also being used, the handheld gamma detector is draped and prepared for use.

Step 3. For RAT mapping, the injection site is surveyed with the gamma detector, and the surgeon then identifies the zone where the radioactivity related to the injection site falls off to background levels. Then the axilla is surveyed, moving the gamma detector slowly. Once a hot spot is identified (counts greater than two-fold background), it must be confirmed that this is not "shine-through" from the injection site by angling the probe tip away from the injection site and pressing down with the probe tip. If a hot spot is present, counts will rise as the probe is pushed closer to it.

Northwestern Handbook of Surgical Procedures, edited by Richard H. Bell, Jr. and Dixon B. Kaufman. ©2005 Landes Bioscience.

Step 4. For RAT lymphatic mapping, the proposed skin incision is marked at the site of the hot spot. If dye tracer alone is being used, the proposed incision is marked at the inferior end of the hair-bearing area of the axilla, along a natural skin crease.

Step 5. The incision is made with a scalpel. It usually does not need to be more than 2 cm long. The dissection is deepened through the subcutaneous fat to the axillary fascia. Blue-stained lymphatic channels encountered superficial to the fascia can be ignored. If RAT mapping is also being used, the probe is inserted into the axillary space and used to direct the dissection towards the hot spot. Again, care is taken not to point the probe towards the injection site.

Step 6. Once a a blue-stained lymphatic channel is identified in the axillary fat, it is followed to a blue-stained lymph node. A node that is significantly replaced by tumor may not stain blue, even though a blue lymphatic channel leads directly to it. If RAT tracing is being used, the dissection is guided by the location of the hot spot. Digital palpation is also extremely helpful in final location of the sentinel node.

Step 7. Once the sentinel node is identified, it is excised by scissor or clamp dissection, applying clips or ligatures as needed for hemostasis. If RAT tracing was used, the excised node is scanned to determine that it is radioactive ex vivo.

Step 8. The axilla is surveyed with the gamma detector to confirm that no additional hot spots are present. If any are identified, they should be excised until the radioactivity of the axilla has been reduced to background levels. If only dye tracer was used, the axillary space is examined for additional blue-stained lymphatic channels or nodes. Any additional such sentinel nodes are excised. Frozen section or touch prep examination of the sentinel nodes is optional, but is advisable if the sentinel node(s) are grossly suspicious, or in the patient who is undergoing mastectomy with immediate reconstruction. If a microscopic diagnosis of sentinel node involvement with tumor is made intraoperatively and the matter has been discussed with the patient preoperatively, the procedure can be converted to a level 1 and 2 axillary dissection.

Step 9. Hemostasis is obtained in the axillary space. The additional infiltration of 0.5% bupivicaine is optional at his point. The wound is closed in two layers, using absorbable interrupted sutures for the subcutaneous tissue and a subcuticular suture for the skin. Steri-strips and a dry dressing are applied.

Postop

Most patients are discharged on the day of surgery. Dressings can be removed and the patient allowed to shower within 1 or 2 days of surgery. Prescription analgesics are required.

Complications

Inability to identify a sentinel node requires conversion to standard axillary dissection. Nerve injury is possible during sentinel node biopsy, particularly to the intercostobrachial nerve. Hematoma and infection are extremely rare.

Follow-Up

Follow-up care may include adjuvant radiation and systemic therapy, as indicated by final pathologic stage of the tumor.

Chapter 63

Partial Mastectomy and Axillary Dissection

Monica Morrow

Indications

The majority of patients with intraductal carcinoma or Stage I and II invasive cancer are candidates for a partial mastectomy. Absolute contraindications to the procedure are multicentric carcinoma, prior irradiation to the breast region, first or second trimester pregnancy, diffuse indeterminate microcalcifications on a mammogram, and the inability to achieve negative margins after an adequate number of surgical attempts. Relative contraindications are a large tumor-to-breast ratio and scleroderma or systemic lupus. Patients with invasive carcinoma who are clinically node negative should have their axillary nodal status determined by sentinel node biopsy or axillary dissection. Those who are clinically node positive require axillary dissection.

Preop

General anesthesia is the method of choice. Long-acting muscle relaxants should be avoided to allow intraoperative nerve identification. Prophylactic antibiotics are used only if a prior surgical breast biopsy has been done and the biopsy cavity will be re-entered or if the patient is immunosuppressed. Deep vein thrombosis prophylaxis with sequential compression devices or subcutaneous heparin is used as indicated by patient risk factors.

Procedure

Step 1. The patient is positioned supine with the ipsilateral arm at nearly a right angle on an armboard. The arm should be circumferentially prepped and draped in stockinette to allow intraoperative movement if needed.

Step 2. The incision is placed over the tumor in the breast. In the superior half of the breast, incisions are in skin creases, at 3 o'clock and 9 o'clock. In the inferior breast, radial incisions are used. Circumareolar incisions should be reserved for tumors in proximity to the areola since they should not encompass more than half the areolar circumference.

Step 3. The subcutaneous fat overlying the tumor is divided and preserved. This will maintain breast contour. Subcutaneous fat needs to be removed only for tumors approaching the dermis.

Step 4. The depth of the tumor is determined by palpation, and the breast tissue overlying the tumor is divided to a depth of 1.0-1.5 cm anterior to the tumor.

Step 5. Double-pronged skin hooks or small Richardson retractors are placed, and using the knife or electrocautery flaps are raised to allow the tumor to be excised surrounded by approximately 1.0-1.5 cm of normal breast tissue. Raising the flaps at the depth of the tumor helps to maintain the breast contour.

Northwestern Handbook of Surgical Procedures, edited by Richard H. Bell, Jr. and Dixon B. Kaufman. ©2005 Landes Bioscience.

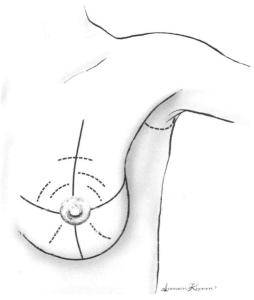

Figure 63.1. Partial mastectomy and axillary dissection.

Step 6. The tumor is controlled and retracted with the nondominant hand while it is sharply excised. This reduces the risk of accidentally cutting into the tumor. Grasping the tumor and surrounding breast tissue with clamps should be avoided as it is difficult to precisely determine the tumor location within the mass of tissue.

Step 7. The specimen is marked with orienting sutures (short suture superior, long suture lateral) and examined to ensure that the tumor is covered on all surfaces by a margin of normal breast tissue. If a margin appears inadequate, an additional specimen is excised with a knife, marked with an orienting suture to indicate the new margin surface, and sent as a separate specimen. Routine frozen sections of margins are not employed.

Step 8. Hemostasis is obtained with cautery, and the walls of the cavity are marked with hemoclips. The wound is packed with a sponge while axillary dissection is carried out.

Step 9. The axillary incision is made at the edge of the hairline in a skin crease and extends from the pectoralis major anteriorly to the latissimus dorsi posteriorly. In patients with a very narrow axillary space, the ends of the incision should be curved superiorly in a U-shaped configuration to provide exposure.

Step 10. Using cautery, the subcutaneous fat is divided. Double-pronged skin hooks are placed and flaps are raised superiorly to the level of the axillary vein, medially to expose the free edge of the pectoralis major, inferiorly to the junction of the breast tissue and the axilla, and laterally to expose the latissimus dorsi.

Step 11. The axillary investing (clavipectoral) fascia is opened medially along the edge of the pectoral muscles, with care being taken not to injure the medial pectoral neurovascular bundle.

Step 12. The latissimus dorsi is cleared of fat along the anterior surface until it turns tendinous. The axillary vein will be identified at this point. Approximately two-thirds of the way to the vein a large branch of the intercostobrachial nerve is encountered. This should be preserved.

Step 13. The axillary vein is cleared of overlying fat from lateral to medial on its anterior surface with care being taken not to strip the vein.

Step 14. The first assistant retracts the axillary contents inferiorly. Dissection 0.5-1.0 cm inferior to the vein is carried out from lateral to medial and venous branches are ligated with 3-0 silk ties.

Step 15. The thoracodorsal bundle is identified and preserved.

Step 16. The axillary fat medial to the thoracodorsal bundle is retracted laterally. Two fingers are inserted into the fat adjacent to the chest wall immediately below the axillary vein and spread in a cranial and caudal direction to expose the long thoracic nerve. The nerve is sharply dissected free from the axillary contents. The medial end of the intercostobrachial nerve is usually identified at this time.

Step 17. The intercostobrachial nerve is dissected free from the specimen and retracted superiorly.

Step 18. The fat between the long thoracic and the thoracodorsal nerves is encircled with a right-angle clamp, divided, and tied proximally. The distal fat is bluntly swept down with a sponge.

Step 19. Small branches of the thoracodorsal vessels entering the specimen are clipped, and the specimen is dissected free of its inferior attachments to the chest wall and sent to pathology.

Step 20. Hemostasis is obtained with clips, ties, and cautery as required. The lumpectomy site is reinspected for hemostasis, and both wounds irrigated with saline. A #19 flat closed suction drain is placed in the axilla.

Step 21. Both wounds are closed with an interrupted layer of 3-0 absorbable sutures, knots inverted, in the deep dermis, and 4-0 subcuticular skin closure. No attempt is made to approximate the cavity in the breast. A light gauze dressing is used.

Postop

Patients are discharged within 24 hours of surgery after instruction in drain care. The drain is removed when the output is less than 30 ml per 24 hours for 2 days. Stretching exercises to maintain shoulder mobility are begun on postoperative day 2.

Complications

Complications include infection, hematoma, seroma, anesthesia in distribution of the intercostobrachial nerve, and damage to thoracodorsal and long thoracic nerves.

Follow-Up

Postoperatively, an examination for arm mobility, seroma formation, and wound healing should be done 1-2 weeks postoperatively. Adjuvant radiation therapy and possibly drug therapy will be required. The patient will require lifelong surveillance for ipsilateral and/or contralateral tumor recurrence.

Chapter 64

Modified Radical Mastectomy

Valerie L. Staradub

Indications

Modified radical mastectomy is indicated for the treatment of invasive breast cancer when breast conserving therapy (BCT) is contraindicated or when the patient chooses mastectomy over BCT.

Preop

General orotracheal anesthesia is indicated. The patient is positioned in the prone position with the arms extended at nearly 90° on armboards. The ipsilateral operative arm is prepped into the field. Standard perioperative antibiotic is administered prior to incision. Long-acting muscle relaxants should be avoided. Deep venous thrombosis prophylaxis with sequential compression devices should be used.

Procedure

Step 1. The subcutaneous tissue of the breast and into the axilla is infiltrated with 500-1500 ml tumescent solution (1 L lactated Ringers' solution with 30 ml of 1% lidocaine solution with epinephrine at 1:1000)

Step 2. An incision is chosen which includes excision of the nipple-areolar complex and incorporates any previous biopsy incision. Remaining skin may be spared if immediate reconstruction is planned, otherwise an elliptical incision is chosen.

Step 3. Skin flaps are raised sharply with a scalpel, extending superiorly to the clavicle, medially to the lateral border of the sternum, inferiorly to the superior aspect of the rectus sheath, and laterally to the latissimus dorsi muscle.

Step 4. The pectoralis major fascia is incised, controlling internal mammary perforators (medially) with ties.

Step 5. The breast and pectoralis fascia are excised with knife or cautery. The breast is left attached inferolaterally to provide traction. The fascia may be preserved in the case of immediate expander reconstruction.

Step 6. The latissimus dorsi muscle edge is followed superiorly along its anterior surface using Richter scissors. Care is taken to preserve the intercostobrachial nerves as encountered. As the muscle becomes tendinous, the axillary vein will be encountered crossing superior to it.

Step 7. The axillary vein is cleared on its anterior surface in a layer-by-layer, lateral-to-medial fashion from the latissimus muscle to the chest wall, taking care not to strip the vein.

Step 8. Dissection is then continued along the axillary vein about 5 mm inferior to the vein, again in a layer-by-layer, lateral-to-medial fashion from the latissimus muscle to the chest wall. There is generally an anterior thoracic branch of the vein that should be controlled, ligated, and divided.

Northwestern Handbook of Surgical Procedures, edited by Richard H. Bell, Jr. and Dixon B. Kaufman. ©2005 Landes Bioscience.

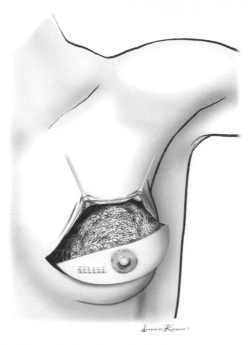

Figure 64.1. Modified radical mastectomy. Incision.

Step 9. The thoracodorsal bundle will come into view during the dissection along the inferior edge of the axillary vein. The nerve is usually slightly posteromedial to the vein. The nerve should be tested with a gentle pinch to confirm contraction of the latissimus dorsi muscle and preserved.

Step 10. Tissues along the chest wall are spread with blunt finger dissection until the long thoracic nerve comes into view. The nerve is tested to confirm contraction of the serratus anterior muscle and preserved.

Step 11. The tissue lying between the nerves is surrounded with a right-angle clamp, taking care not to catch the nerves in the clamp. The tissue is divided and tied at its superior aspect to control lymphatics and small vessels.

Step 12. The tissue remaining between the nerves is swept inferiorly with an open gauze sponge over the first two fingers.

Step 13. The intercostobrachial nerves are separated from the specimen and retracted out of the way.

Step 14. The thoracodorsal bundle is followed inferiorly along its anterior surface, clipping and dividing small vein branches as they arise.

Step 15. When the thoracodorsal bundle turns and dives into the latissimus dorsi muscle and the long thoracic nerve turns and enters the chest wall, the remainder of the specimen is removed by electrocautery where it remains attached inferolaterally, doing so in such a way that the breast and axillary portions of the specimen remain intact.

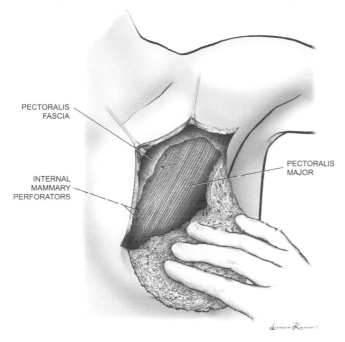

Figure 64.2. Modified radical mastectomy. Extent of depth of dissection.

Step 16. A #19 round Jackson-Pratt drain is placed beneath the inferior skin flap and brought out through a separate stab incision. A second drain is placed into the axilla, taking care that it does not rub on the axillary vein. The drains are secured to the skin with nylon drain stitches.

Step 17. The deep dermal layer is closed with interrupted 3-0 absorbable sutures and the skin with running subcuticular 4-0 absorbable suture.

Postop

The Jackson-Pratt drains should be stripped periodically and output recorded. The drains can be removed when output is less than 30 cc/day for 2 days. Range-of-motion exercises should be initiated about 2 days after surgery.

64

Complications

The most common complications associated with this procedure are hematoma beneath flaps, skin necrosis, lymphedema, and/or arm numbness.

Follow-Up

The patient is referred for chemotherapy, endocrine therapy, and/or radiation therapy as needed according to the stage of tumor. The patient should be monitored for local recurrence and have annual examinations and mammography to look for primary disease in the contralateral breast.

Chapter 65

Simple Mastectomy

Valerie L. Staradub

Indications

Simple mastectomy is indicated for extensive ductal carcinoma in situ (DCIS) of the breast, DCIS in a patient who is not a candidate for radiation therapy, or DCIS in a patient who elects mastectomy over breast conservation therapy. Simple mastectomy may be combined with sentinel node biopsy in selected cases of invasive carcinoma.

Preop

General orotracheal anesthesia is preferred. Intravenous access should be on the nonsurgical side. Prophylactic antibiotic is given 30 minutes prior to incision. Deep venous thrombosis prophylaxis with sequential compression devices and/or subcutaneous heparin is used depending on patient risk factors.

Procedure

Step 1. The patient is placed in prone position with the arms extended nearly 90° on armboards. The subcutaneous tissue of the breast is infiltrated with 500-1500 ml of tumescent solution (1 L lactated Ringers' solution with 30 ml of 1% lidocaine solution with epinephrine at 1:1000).

Step 2. An elliptical incision is made to include excision of the nipple-areolar complex and any recent biopsy incision. Other skin may be preserved if immediate reconstruction is planned, otherwise excess breast skin is excised.

Step 3. Skin flaps are raised sharply with a knife superiorly to the clavicle, medially to the lateral border of the sternum, inferiorly to the superior aspect of the rectus sheath, and laterally to the latissimus dorsi muscle.

Step 4. The pectoralis major fascia is incised, controlling internal mammary perforators (medially) with ties.

Step 5. The breast and pectoralis fascia are removed from superior to inferior with knife or cautery. Fascia may be preserved if immediate expander reconstruction is planned.

Step 6. Laterally, the latissimus dorsi muscle is followed superiorly until the axillary investing fascia is entered, to ensure removal of the axillary tail of breast tissue. The axillary tail is marked with a stitch for orientation of the specimen.

Step 7. A #19 round closed Jackson-Pratt suction drain is placed beneath the inferior skin flap, brought out through a separate stab incision, and secured with a nylon drain stitch.

Step 8. The deep dermal layer is closed with interrupted 3-0 absorbable sutures. The skin is closed with a running subcuticular 4-0 suture.

Northwestern Handbook of Surgical Procedures, edited by Richard H. Bell, Jr. and Dixon B. Kaufman. ©2005 Landes Bioscience.

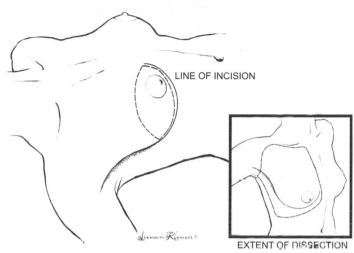

LINE OF INCISION

EXTENT OF DISSECTION

Figure 65.1. Simple mastectomy.

Postop

The Jackson-Pratt drain should be stripped perioidically and output recorded. The drain can be removed when output is less than 30 cc/day for 2 days.

Complications

Most common complications include hematoma beneath the flap and skin necrosis.

Follow-Up

The patient should be monitored indefinitely for local recurrence or contralateral disease. Tamoxifen therapy for contralateral risk reduction should be considered.

65

Chapter 66

Major Excision and Repair/Graft for Skin Neoplasms

Jeffrey D. Wayne

Indications

Until recently, all but the thinnest of invasive melanomas of the trunk and proximal extremities were treated with radical resections, with 3-5 cm margins. While rates of local recurrence were low, most of these excisions required a skin graft for closure. Four randomized, prospective surgical trials including the WHO and Intergroup trials challenged such wide margins of excision and established the treatment standards employed in the management of malignant melanoma today. Namely, in situ lesions are routinely resected with a margin of 0.5-1.0 cm. Thin melanomas, with a Breslow's depth of less than or equal to 1.0 mm, are excised with 1 cm margins. Intermediate-thickness lesions can be safely excised in all locations with a 2 cm margin. Finally, while they may have a higher propensity to local recurrence, approaching 11%, thick (> 4.0 mm) lesions are also excised with a 2.0 cm margin, as these patients are more likely to succumb from distant disease.

Preop

Most often, the primary has been resected either by a punch or excisional biopsy. For lesions on the trunk, or proximal extremity, primary closure is the norm. Most patients can undergo elliptical excision of the scar and tumor bed under local anesthetic, with or without intravenous sedation. In general, a proper excision involves resection of the scar and tumor bed, with an appropriate margin of normal tissue, and an en bloc resection of the underlying subcutaneous tissues down to the underlying muscular fascia. There is no evidence that removal of the underlying fascia improves local control or survival, and thus it is usually spared.

This procedure is usually performed on an outpatient basis, unless a skin graft is to be performed. Either way, the patient is admitted to the operating room on the morning of surgery. If intravenous sedation or a general anesthetic is to be used, the patient is instructed to take nothing by mouth after midnight. If a previous biopsy has been performed, a single dose of a first-generation cephalosporin is given in the holding area, at least 45-60 minutes prior to the planned incision. If the patient has an allergy to penicillin, clindamycin may be used as an alternative.

Procedure

Step 1. To start the procedure, the patient is prepped and draped in the usual sterile fashion. A sterile ruler is then used to measure the desired (0.5-2.0 cm) margin from all edges of the primary lesion or biopsy scar. This usually results in a circular or oval area to be excised. The incision is extended in either direction to

---

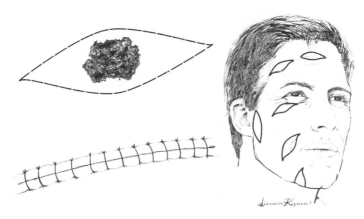

Figure 66.1. Major excision and repair for skin neoplasms.

form an ellipse. In general the ellipse should be oriented along Langer's lines, or as necessary to allow for maximal utilization of available local skin. To achieve a closure with minimal tension, the length of the incision must be at least three times the measured width of the planned excision. Local anesthetic (usually 1% lidocaine with or without epinephrine, mixed in a 1:1 concentration with 0.5% bupivicaine) is then instilled into the area to be resected. A number 15 blade is used to create the incision. The incision is carried down to the level of the underlying muscular fascia, in a perpendicular fashion, using electrocautery. Once the boundaries of the excision have been defined, the specimen is dissected off the underlying fascia using electrocautery.

Step 2. After the specimen has been excised, it is oriented for the pathologist with a marking stitch of 2-0 or 3-0 silk and sent for permanent sections. The wound is then irrigated with sterile saline, and a meticulous check for hemostasis is undertaken. Final hemostasis is usually achieved with electrocautery, and/or 3-0 undyed, absorbable ties. Single skin hooks are then used to apply slight traction to ends of the incision, and a marking pen is used to place lines across the incision to help with suture alignment. Alternatively, a series of towel clips can be used to align the wound for closure. Depending on the size of the resultant soft-tissue defect, a closed-suction drain may be placed in the wound bed prior to closure. The deep dermis is then reapproximated with an interrupted layer of 3-0 undyed, absorbable sutures, placed no further than 5 mm apart. The skin may be closed with either interrupted 3-0 nylon sutures or a running 4-0 monofilament subcuticular stitch, depending on the tension on, anatomic location of, and anticipated mobility of the individual incision.

Step 3. If the incision will not close in a primary fashion, local advancement flaps may often be used. Alternatively, a split-thickness skin graft may be employed. For lesions on the extremities, the donor site is usually chosen on the opposite extremity. A Zimmer dermatome is used to harvest a graft of 0.16 mm in thickness. A mesher or scalpel is then used to create holes in the graft and allow for the seepage of serum from the wound. Should a seroma or hematoma form under the graft, the take will be poor.

66

The wound bed is prepared by imbricating the edges of the wound with absorbable sutures. This minimizes the area to be covered and allows for a smooth transition from the wound to the surrounding skin. The graft is secured in place with 4-0 absorbable sutures and then a nonadherent (petroleum gauze) dressing is placed over the graft. A bolster of cotton balls soaked in glycerin is then placed over the petroleum gauze dressing and held in place using 3-0 silk sutures placed at the periphery of the wound. The donor site may be dressed with petroleum gauze or simply a Tegaderm or other occlusive dressing. The area is immobilized for approximately 7 days, at which time the dressings are taken down and take of the graft is assessed.

Postop

As above, most patients are discharged to home on the day of surgery. If a closed suction drain has been placed, the patient is instructed on care of the drain, asked to empty the drain twice daily, and to record the output. The patient is brought back to the clinic for drain removal when the total output for 24 hours is less than 30 cc for 2 days in a row. This is usually the case by postoperative day 5. Sutures are removed 10-14 days after the operation and steri-strips are placed.

Complications

Cellulitis or minor wound infections are the most common complications. The first line of treatment is often oral or intravenous antibiotics, with elevation of the extremity in question. Only rarely will sutures have to be removed and the wound allowed to close by secondary intention. Wound-edge separation with the formation of a wide scar is another complication. This often results from the use of a subcuticular closure in a high-tension wound or from early suture removal.

Follow-Up

Current guidelines call for all patients to be examined, with a full head-to-toe skin survey and palpation of all lymph node basins, at least twice a year. Many physicians who care for patients with melanoma will also follow chest X-rays and LDH levels; however, this is considered optional. Any new symptoms should prompt radiologic imaging as appropriate.

Chapter 67

Sentinel Lymph Node Biopsy for Melanoma

Jeffrey D. Wayne

Indications

Although therapeutic lymph node dissection remains the procedure of choice for grossly involved regional lymph nodes in melanoma patients, elective lymph node dissection (ELND) has been abandoned as treatment for the clinically negative lymph node basin. Sentinel lymph node biopsy (SLNB) is a more accurate and less morbid way to assess the regional lymphatics in patients with negative physical findings. The operative concept is that individual areas of skin have specific patterns of drainage, not only to a regional basin, but to a specific node or nodes within that basin. This sentinel lymph node may be identified at the time of operation using a vital blue dye, a radiolabeled colloid, or both.

Thin melanomas (Breslow's depth < 1.0 mm) have a low propensity to metastasize to regional lymph nodes (<5%). Thus, patients with thin lesions are usually not offered a SLNB, unless they have a lesion with one or more adverse prognostic indicators. Such factors are: Clark's level > III, ulceration, evidence of regression, or a prominent vertical growth phase. All patients with intermediate thickness and thick lesions, with clinically negative nodes (clinical Stage I and II), are appropriate candidates for SLNB.

Preop

Successful SLNB is predicated upon identifying the lymph node basin(s) at risk. While this is a relatively straightforward task in extremity lesions, there is often ambiguous drainage from lesions located on the trunk, head, and neck. Thus, preoperative lymphoscintigraphy is crucial to operative planning. Furthermore, preoperative lymphoscintography helps identify not only the number and relative location of the sentinel lymph nodes within a particular basin, but it also allows for the identification of "in transit" sentinel lymph nodes. These nodes, often found in the epitrochlear or popliteal spaces, may be sources of regional or distant failure if not recognized and properly excised.

The highest reported rates (in excess of 97%) are found when both 1% isosulfan blue dye and ^{99m}technetium sulfur colloid are used for localization. Sentinel lymph node biopsy is ideally performed at the same operative setting as excision of the primary lesion. It is an outpatient procedure, and patients can almost always be discharged within 23 hours of operation.

The procedure may be done under local anesthesia with intravenous sedation or under general anesthesia. The patients are seen in the nuclear medicine department upon arrival to the hospital, and the preoperative lymphoscintigraphy is performed. A four-point intradermal injection technique is used to administer 0.5-1 mCi of

Northwestern Handbook of Surgical Procedures, edited by Richard H. Bell, Jr. and Dixon B. Kaufman. ©2005 Landes Bioscience.

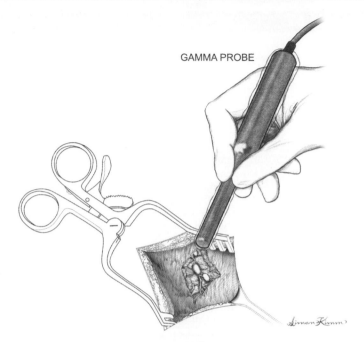

GAMMA PROBE

Figure 67.1. Sentinel lymph node biopsy for melanoma.

^{99m}technetium sulfur colloid. Early and frequent images using a gamma camera are obtained in anterior and lateral views until all sentinel lymph nodes are identified.

The patient is then transported to the preoperative holding area, where a hand-held gamma probe is used to verify the location transcutaneously.

Procedure

Step 1. The patient is positioned so as to allow access to the primary site and any lymph node basins harboring a sentinel node. This sometimes requires the use of a beanbag and the lateral decubitus position. Prior to prepping, 1-3 ml of 1% isosulfan blue dye is injected sterilely in the area around the primary lesion or biopsy scar. The patient is then prepped and draped in the usual sterile fashion.

Step 2. Attention is first turned to the sentinel lymph node basin. A handheld gamma probe is brought on the field in a sterile sheath and used to localize the node in question. A 15 blade is then used to make a focused 2-3 cm incision directly over the highest point of radioactivity. Care must be taken to align the incision such that it can be easily excised should a completion lymph node dissection be necessary.

Step 3. A self-retaining retractor may then be placed to facilitate dissection. Careful blunt and sharp dissection with meticulous hemostasis are then used until afferent and efferent blue lymphatic channels are identified. These channels are traced to the first blue node which is isolated and excised. Afferent and efferent channels as well as small vessels may be controlled using 3-0 or 4-0 absorbable ties. The use of hemoclips is discouraged, as they may interfere with a completion dissection. The sentinel lymph node is then scanned ex vivo, and the maximal count is recorded. The basin is then rescanned, and any other blue nodes or any node with an in vivo count within 10% of the hottest node should be removed. Nodes are individually identified and sent to pathology for permanent, serial sectioning histology and immunohistochemistry.

Step 4. Once the background count in the basin is less than 10% of the hottest node, a sterile moist gauze is placed in the wound, and attention is turned to the primary lesion. Using a second set of instruments, a wide local excision and primary closure is performed, as described previously. Gloves and instruments are again changed, and the lymph node basin and all intervening tissues are rescanned to rule out any additional in transit nodes. The sentinel node wound is then closed in two layers using an absorbable suture to close the deep dermis and a monofilament suture to close the skin in a subcuticular fashion. Steri-strips and a dry sterile dressing are then placed.

Postop
Patients may be discharged on the day of surgery with oral analgesics.

Complications
Unlike with ELND or therapeutic lymph node dissection, complication rates with SLNB are low. The most common complications are cellulitis, minor wound infections, and seroma formation. Large seromas may be drained in the office using a sterile technique and 18-gauge needle. Lymphedema is rarely observed.

Follow-Up
Should a positive sentinel lymph node be found, completion lymph node dissection is usually scheduled within 2-3 weeks. The patient is then referred to a medical oncologist for consideration of treatment with interferon α-2β or biochemotherapy.

Chapter 68

Radical Excision of Soft Tissue Tumor (Sarcoma)

Jeffrey D. Wayne

Indications

Soft tissue sarcomas (STS) are a rare and diverse group of tumors. It is estimated that there will be over 8,000 new cases diagnosed, and almost 4,000 deaths from STS in the United States in 2004. Most of these tumors are derived from the embryonic mesoderm or ectoderm, and individual sarcomas take their name from the parent tissue of origin. The mainstay of treatment for most sarcomas is surgical. However, in an effort to facilitate limb salvage, radiation therapy is frequently employed in either the neoadjuvant or adjuvant setting. This is especially true if the individual lesion is high grade, large (T2 or $\geq$ 5 cm in greatest dimension), or deep. A typical course consists of 50 Gy given over 25 fractions. Sarcomas of the extremity typically present as a painless mass. An MRI or CT scan of the affected limb usually suggests the diagnosis. However, a confirmatory biopsy is often required, especially if a neoadjuvant treatment strategy is to be employed. We advocate the liberal use of core-needle biopsy. This is an office-based procedure, which may be accomplished under a local anesthetic. Deeper lesions may undergo core biopsy in interventional radiology under ultrasound or CT guidance. Rarely, an open (incisional) biopsy is required. If this technique is employed, care must be taken to orient the incision along the long axis of the limb so as to facilitate future attempts at limb salvage. To illustrate some of the concepts of radical excision for STS, we will outline the en bloc excision of a sarcoma of the anterior compartment of the thigh.

Preop

Testing is dictated by the location, size, and grade of the sarcoma. For a small, superficial (T1, $\leq$ 5cm) sarcoma a chest X-ray and CBC may be all that is required. For larger (T2, > 5cm) or high-grade lesions CT scan of the chest, and abdomen and pelvis (if a lower extremity sarcoma), are routinely obtained to rule out disseminated disease. If a previous biopsy has been performed, I typically give a single dose of a first-generation cephalosporin in the holding area. If the patient is penicillin allergic, clindamycin may be substituted.

Procedure

Step 1. The patient is taken to the operating suite and placed in the supine position with the affected limb slightly abducted and externally rotated. A sterile bump of folded towels provides adequate support beneath the knee. Care must also be taken to provide adequate padding under the lateral malleolus. A general anesthetic is usually employed, as an epidural catheter may interfere with neurologic monitoring of the extremity. We also typically ask the anesthesiologist to avoid the administration of neuromuscular blocking agents. The limb is prepped from the knee to a level above the anterior superior iliac spine. A circumferential prep is often helpful to allow for repositioning of the leg as necessary.

Step 2. The operation begins by making an elliptical incision around any previous biopsy incision or scar, with 1-2 cm margins. This incision is extended proximally and distally over the course of the longitudinal extent of the lesion. Medial and lateral flaps are then raised with the assistance of skin rakes.

Step 3. In the case of large tumors, it is often helpful to expose major anatomic structures, such as vessels and nerves proximal and distal to the lesion. This will facilitate their salvage if oncologically feasible. Thus, in the case of a deep thigh sarcoma, the common femoral artery, vein, and nerve are identified cephalad to the tumor, and encircled with vessel loops. Similarly, the superficial femoral vessels are located at the medial aspect of the distal thigh and likewise encircled with vessel loops.

Step 4. Deep dissection is then begun at whichever aspect of the mass that allows for maximal dissection with the least amount of resistance. Care should be take to stay outside the tumor "capsule" which is a misnomer in most cases of sarcoma. This is actually a pseudocapsule composed of compressed tumor cells at the periphery of the mass. A bovie electocautery, 10 blade, or Metzenbaum scissors are used as appropriate for the thickness and location of the tissue. Any muscle fibers, for example from the rectus femoris, adductor magnus, or vastus medialis, are taken en bloc with the specimen. Only rarely must an entire muscle belly be sacrificed in an anatomic fashion.

Step 5. The deep aspect of the tumor is addressed last. If the mass can be removed in a margin-negative fashion without sacrifice of the vessels then this is optimal. However, vascular reconstruction with reversed saphenous vein or polytetrafluoroethylene (PTFE) is an option, if necessary. If a positive margin is unavoidable in the vicinity of bone, or a major motor nerve, thought should be given to the placement of brachytherapy (after loading) catheters in the wound bed. In any instance, it is always advisable to place large, titanium hemoclips in the resection bed to facilitate adjuvant external-beam radiation.

Step 6. Finally, the wound is irrigated with copious amounts of warm saline solution and a meticulous check of hemostasis is made. A flat, closed-suction drain is placed in the resection bed and brought out through an inferiorly placed stab incision, in line with the primary incision. The deep dermis is then reapproximated with a layer of 3-0 absorbable sutures in an airtight fashion and then the skin is closed with interrupted nylon sutures in an interrupted simple or vertical mattress fashion.

Postop

In the absence of vascular or plastic surgery reconstruction, the patient is typically kept in the hospital overnight. The leg is elevated, and the patient is instructed in drain management. They are asked to return to the clinic when the drain output is less than 30 cc for two consecutive 24-hour periods. The patient is seen by the physical therapist on postoperative day one for instruction in crutch walking. The patient is given a prescription for an oral narcotic to take home (e.g., Vicodin). Sutures are removed at 2 weeks.

68

Complications

Inadvertent nervous or vascular injury may be minimized by utilizing the techniques outlined above. Specifically, dissecting out major anatomic structures prior to any attempt at removal of the main tumor mass will allow for these structures to be followed along their course and preserved, as they are dissected sharply away from the mass. Wound infection rates in the range of 25-30% are reported when external beam radiation is employed in a neoadjuvant fashion. Liberal use of autologous rotation and free flaps are advised in this instance.

Follow-Up

Patients are seen at intervals of 4-6 months for the first 2-3 years. At each visit a physical exam, chest X-ray, and site-specific imaging (CT, MRI, or ultrasound) is performed. If the lesion was large and high grade, a CT scan of the chest is substituted for the chest X-ray, due to the high risk of pulmonary metastases. Follow-up is continued for 10 years due to the small, but definable instance of late recurrence.

Section 4: Plastic Surgery
Section Editor: Thomas Mustoe

Chapter 69

Burn Debridement and/or Grafting

Julius W. Few, Jr.

Indications

First degree—Superficial partial-thickness burns. Blisters with pain. Superficial epidermal burns, such as from sunburn, heal with topical therapy alone.

Second degree—Deep partial-thickness burns. Less pain, deep dermis exposure with or without hair follicle exposure.

Third degree—Full-thickness burns. No pain, dry, leather-like appearance.

Fourth degree—Muscle or bone exposure.

Preop

The surgeon must assess and manage (A)irway, (B)reathing, and (C)irculation, as with any form of trauma. Estimate the total body surface area (TBSA) burn using the rule of nines. Intravenous fluid replacement is done using the Parkland formula, 4 cc lactated Ringers/kg body wt/%TBSA burn, given over the first 24 hours: 1/2 in the first 8 hours. Continue IV fluids to maintain a urine output of 0.5 cc/kg wt/ hours. Avoid systemic antibiotics. Topical antimicrobial agents—silvadene, sulfamyalon, or silver nitrate—are used liberally. Silvadene is the agent of first choice; silver nitrate is used in patients with sulfa allergy; and sulfamyalon is used for patients who require eschar penetration. Treat associated trauma.

Procedure

Step 1. Bring patient to room with temperature greater than 75° F.

Step 2. Anesthesia with appropriate monitoring.

Step 3. Prep and drape all involved areas.

Step 4. Debride all nonviable tissue. Tangential excision is preferred using a burn knife or dermatome.

Step 5. Avoid skin grafting directly onto adipose tissue.

Step 6. Using a dermatome (Zimmer), obtain a skin graft from an uninjured area. Set to thickness of 12-15/1000 in.

Step 7. Consider meshing skin 2-3:1 to increase surface area coverage while minimizing the donor area. Avoid meshed grafts for the hand.

Step 8. Hemostasis using topical thrombin and cautery.

Step 9. Apply meshed skin graft with staples or sutures.

Step 10. Cover skin graft with adaptic or Owens gauze to avoid skin graft disruption from the removal of the dressings.

Step 11. Apply mineral oil-soaked cotton or gauze to the adaptic dressing.

Step 12. Apply Reston foam dressing with staples or Ace wrap to hold dressing in place.

Northwestern Handbook of Surgical Procedures, edited by Richard H. Bell, Jr. and Dixon B. Kaufman. ©2005 Landes Bioscience.

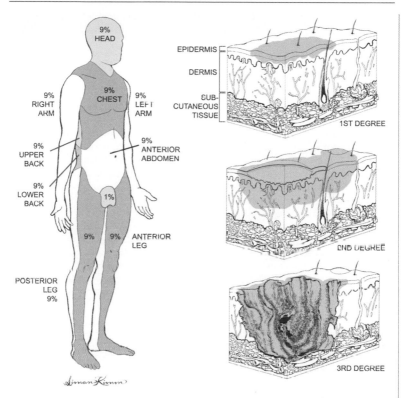

Figure 69.1. Burn debridement and/or grafting.

Step 13. Cover donor site with Biobrane™ or Opsite™ dressing.
Step 14. Do not graft more than 25% TBSA in one sitting.

Postop

Control temperature. The dressings are kept in place for 3-5 days. Fluid and nutritional support. Begin physical therapy at 7-10 days.

Complications

Hematoma/seroma, infection, graft failure, scarring/deformity, and need for additional surgery.

Follow-Up

Aggressive physical therapy. Compressive garments. Secondary reconstruction. Scar/deformity management.

Chapter 70

Split-Thickness Skin Grafts

Julius W. Few, Jr.

Indications

Split-thickness skin closure is useful for wounds with a stable granulating or vascular base for plasma diffusion/neovascular growth. Split-thickness skin grafts (STSG) are very reliable but require a donor area that will heal by secondary means. STSGs tend to contract over time and should not be used in wounds which will not tolerate contracture (i.e., hand).

Preop

The desired skin graft site should have a stable, clean base of granulation tissue or muscle/fascia to allow for successful plasma diffusion/neovascularization. Wound biopsy should be done if the wound is suspected to have greater than 500,000 microorganisms. Repeated wound debridement may be necessary to achieve a clean base for grafting.

Procedure

Step 1. The patient is positioned to allow access to the wound (recipient site) and the donor site.

Step 2. The donor area (typically the lateral thigh) is prepared by removing betadine and applying mineral oil to reduce dermatome friction.

Step 3. A hydraulic dermatome or knife is set 12-16/1000th inch thickness, depending on wound location and/or type.

Step 4. Using four-point traction on the donor area, the dermatome is placed at a 45° angle and advanced with uniform pressure to harvest desired skin.

Step 5. The skin graft can be meshed to increase surface area and/or to permit fluid egress, which may hinder graft take.

Step 6. The wound (recipient site) is cleaned/debrided to healthy, bleeding tissue.

Step 7. The skin graft is applied to the wound, dermis side down, and secured with sutures or staples. ALWAYS CHECK DERMIS SIDE DOWN.

Step 8. Adaptic™ or Owens gauze is applied to the secure skin graft to prevent skin adhesion to the dressing.

Step 9. Mineral oil-soaked cotton balls or Reston foam is applied to the adaptic.

Step 10. A second Reston foam pad is sewn or stapled to keep shear movement to a minimum, with moderate compression. Ace wrap can be used on the extremities.

Step 11. The donor site is checked for hemostasis, then a Tegaderm™ or Biobrane™ dressing is applied.

Northwestern Handbook of Surgical Procedures, edited by Richard H. Bell, Jr. and Dixon B. Kaufman. ©2005 Landes Bioscience.

Postop

Prophylactic antibiotics may be used. The area should be elevated and kept dry for 3-7 days. The recipient dressing is removed at 3-7 days. The donor dressing is left on until it falls off or shows signs of delayed healing. The donor area should re-epithelialize by 7-14 days. Activity should be limited while the skin graft dressing is on.

Complications

Bleeding, infection, neurovascular injury, delayed wound healing, skin graft loss, scarring, and need for additional surgery.

Follow-Up

Once the grafted area is exposed, it should be kept clean and lubricated. After washing, aloe or vitamin A and D ointment can be used daily. Split-thickness skin grafts lack dermal support structures and are at risk for dessication.

70

Chapter 71

Debride/Suture Major Peripheral Wounds

Gregory Dumanian

Indications

Major acute wounds of the head and neck, chest, abdomen, back, and extremities should be closed promptly to restore patient homeostasis. Chronic wounds, in contradistinction, should be assessed for etiology, reasons for failure to close, and appropriateness for closure.

Preop

As all major acute wounds should be closed, all that is left to decide is the method of wound closure and its timing. The method of wound closure depends on the importance of the soft tissues exposed. Exposed fractures and orthopedic hardware, tendons, open joints, recent incision lines in blood vessels or bowel, and prosthetic material all require soft tissue flaps to aid in wound healing. Surface wounds without critical underlying exposed tissue often heal with split-thickness skin grafts. Timing of major reconstructive surgery is an interplay between overall patient status, the ability of the underlying tissue to remain "open," and the complexity of the reconstructive procedure. For extremity wounds, assessment of distal limb nerve, artery, and muscle function is critical.

Procedure

Step 1. Patient positioning should be appropriate for exposing the major wound and any other incision for flap and skin graft harvest. The greater the patient exposure, the better.

Step 2. "Create the defect" by first sharply debriding devitalized tissue. Aim to cut back to normal-appearing tissue in order to reduce bacterial loads and to allow for primary wound healing.

Step 3. Pulse-lavage tissues which are critical, contaminated, and cannot be removed during the sharp debridement process.

Step 4. Assess the wound and the ability of tissues to either be closed primarily, to receive a skin graft, or those wounds which will need flap coverage.

Step 5. Wounds which can be closed primarily should not have undue tension, should not require more than a 2-0 nylon to appose the tissue, and should not put undue pressure on underlying tissue. Drain the underlying soft tissues liberally. Keep deep foreign material in the wound to a minimum to help avoid postoperative infections.

Step 6. Split-thickness skin grafts are best harvested from the posterolateral thigh with a mechanical dermatome at 12-14/1000th inch. Donor sites are dressed with a semipermeable membrane dressing. The grafts are usually meshed and held in place on the wound with stent dressings.

Northwestern Handbook of Surgical Procedures, edited by Richard H. Bell, Jr. and Dixon B. Kaufman. ©2005 Landes Bioscience.

Step 7. Pedicled flaps from adjacent tissue require a knowledge of flap and blood vessel anatomy. Ensure that the flap blood vessels have not been damaged by the agent which caused the wound. "Creating the defect" first before flap elevation ensures proper flap dimensions for wound coverage. Elevate the flap, close the donor site, and then insert the flap on the wound. Liberally drain the donor site and the flap recipient site.

Step 8. For major wounds requiring free flap transfer, the flap donor site will be removed spatially from the wound. Blood vessels near the wound will need to be dissected out for eventual anastomosis to the flap blood vessels. Donor site closure and flap inset will be the same as for pedicled flaps.

Step 9. Immobilization of major wounds with postoperative dressings is a critical feature of these procedures. Plaster splints made in the operating room are often used to immobilize soft tissues and help to prevent unwanted movement and potential late joint contractures.

Postop

Elevation, immobilization, and pressure relief of recently treated wounds will decrease edema and help wound healing. Skin graft dressings can be removed 3-5 days after placement. Flaps need to be checked for vascularity.

Complications

Wound dehiscence, wound infection, hematoma, partial or total skin graft loss, partial or total pedicled flap loss, free flap arterial or venous thrombosis.

Follow-Up

Initial acute wound management may only be the first phase in eventual reconstruction. Extremity injuries frequently require physical or occupational therapy to restore function.

71

Chapter 72

Repairing Minor Wounds

Thomas Mustoe

Indications

Indications include lacerations penetrating the superficial dermis with some gaping of the wound; blunt trauma injuries causing splitting of the skin with some gaping; and combination laceration/abrasion injuries where skin can be reapproximated even in the setting of surrounding abrasions.

Preop

Anesthetize the area if sensitive to allow proper cleaning. Utilize xylocaine with epinephrine in the head and neck to allow sufficient anesthesia time to complete the repair and minimize bleeding. Avoid epinephrine in the hands or feet, particularly in the digits. Arterial spasm with eventual digit necrosis is a well-known complication of utilizing epinephrine in those regions. For young children consider an oral or intramuscular sedative, although it may not be necessary. Sedatives can make the experience less traumatic and allow a more meticulous repair. For infants, immobilize the baby in a papoose or blanket. Prep the area with betadine and utilize a simple drape. Two towels work fine or optimally a single sheet with a cut-out circle.

Procedure

Step 1. Irrigate the wound thoroughly. Irrigation under adequate pressure is much more effective than soaking or low-pressure irrigation. Utilizing a 10 cc syringe with a 20-gauge needle at close range will deliver adequate force. For a small wound 30-40 cc of saline should suffice unless the wound is very dirty and there is ground-in foreign material. With abrasions, it is critical to remove all foreign material to prevent tattooing. That may necessitate extensive irrigation.

Step 2. Excise ragged or extensively traumatized wound edges, as long as the wound can be brought together with minimal tension. The eventual scar will be superior. An easy method is to score the 1-2 mm excision at the edge and then complete the excision of the traumatized edge with scissors.

Step 3. If the wound is gaping substantially and penetrates the entire dermis, utilize buried intradermal sutures with the knot on the deep surface of the tied ligature. Absorbable sutures such as polygalactic acid should be utilized.

Step 4. For optimal epidermal approximation, cutaneous sutures with either a permanent suture or an absorbable suture can be utilized. In the head and neck a permanent suture such as 6-0 or 5-0 nylon works well. Alternatively, for children or when follow-up is uncertain, a 5-0 fast-break plain gut or a 6-0 mild chromic suture can be utilized. The disadvantage of absorbable sutures is an increased risk of inflammation with potential deleterious effects on eventual scar outcome. In general interrupted sutures are safer in the case of infection in a dirty wound, but

Northwestern Handbook of Surgical Procedures, edited by Richard H. Bell, Jr. and Dixon B. Kaufman. ©2005 Landes Bioscience.

in a clean wound with good dermal approximation, a running suture is very acceptable and saves time. In the hands and feet, a one-layer closure with 5-0 nylon is preferred.

Step 5. Occlusive dressings minimize inflammation and optimize the eventual scar outcome. In routine cases this is best accomplished by steri-strips and an adhesive such as benzoin or gum mastic (Mastisol). If significant exudate is expected, or there are surrounding abrasions, then an occlusive ointment applied repeatedly is a good alternative. Antibiotic ointments, particularly those that contain neomycin, run the risk of allergy. If the wounds are kept clean of exudate in the postoperative period, then a non-antibiotic ointment such as A & D ointment will minimize the risk of allergic dermatitis.

Postop

The major principle to recognize is that bacterial growth in the eschar of and serum on the top of an incision will lead to inflammation, delayed epithelization, and a suboptimal scar. Therefore if the wound is clean and postoperative exudate is presumed to be minimal, then steri-strips will protect the wound and prevent bacterial overgrowth. There is no need to keep the area dry in the shower. If the wound is dressed without an occlusive dressing, ointment should be applied and the area should be washed free of dried blood and protein exudate at least daily to prevent bacterial overgrowth in the microenvironment of the wound, with application of an appropriate ointment as described above.

Complications

If the wound has been properly irrigated, with removal of foreign body and devitalized tissue, wound infection should be very unusual (1-3%). If interrupted sutures are used, then a suture can be removed to allow drainage of a superficial abscess in the postoperative period.

Follow-Up

In cosmetically sensitive areas such as the face, neck, or upper chest, periodic follow-up is necessary to observe for the possibility of hypertrophic scar or keloid. If poor scarring develops, intralesional steroids or topical silicone gel sheeting should be instituted. If these steps are not successful or if the physician is inexperienced with these methods, appropriate referral to a plastic surgeon should be considered.

Chapter 73

Removal of Moles and Small Skin Tumors

Neil A. Fine

Indications

Growing lesions, changing lesions, painful or bleeding lesions, darkening of pigmentation, elevation of lesion, or patient preference, with the knowledge that there will be a scar.

Preop

This is typically an office-based procedure but may be done under local or in an ambulatory operating room setting, or as part of a more significant operation. It can be prudent to stop aspirin and anticoagulants, but often these procedures are so minor that even this is not necessary.

Procedure

Step 1. Prep the skin with alcohol.

Step 2. Mark the lesion for excision along the skin tension lines. Elliptical excisions are usually preferred. When in doubt, excise the lesion as a circle and determine later which orientation to close.

Step 3. Inject with lidocaine and epinephrine.

Step 4. Allow 5-10 minutes to pass between injection and incision.

Step 5. Prep with betadine and sterile towels.

Step 6. Excise the lesion with a 15 blade. Extend into the subcutaneous tissue. Include a margin of 2-3 mm on suspicion of skin cancers. On lesions for which you have not previously decided upon orientation, a circular excision will have been done and at this stage pull gently with tooth forceps to see which direction the skin closes most easily. Once this is determined, place a deep buried suture in the middle, and then remove the resultant dog ears as necessary.

Step 7. Finish closing with further buried sutures and if necessary external sutures; then cover with skin tape. Internal buried sutures will take tension off the external closure and may lead to a finer scar.

Step 8. Send all suspected skin cancers and all pigmented lesions for pathology.

Additional Comments

Many variations of suture techniques exist. My preference is to use internal clear nylon sutures to provide long-lasting support to the scar. External clear nylon sutures are then used as necessary to get adequate skin coaptation. Using a single suture for both the internal and external suture cuts down on the number of sutures that are necessary for these minor procedures. I typically will use a 5-0 suture on the face and 4-0 sutures elsewhere on the body. I have found that a single layer of 4-0 polypropylene typically works well on the scalp, and the blue

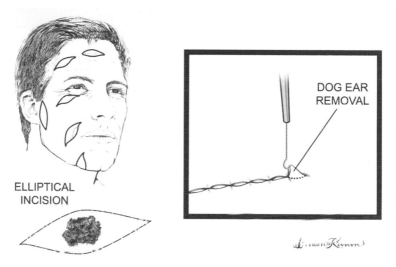

Figure 73.1. Removal of skin moles, small tumors, etc.

color allows for easier identification. A small disposable stapler also works well in the scalp. For postoperative care, pressure is held on the excision for 15-20 minutes to reduce bruising. Patients may shower the next day.

Complications

Bleeding and infection are always possible with any excision, but they are quite rare with simple skin excisions. They are especially rare in the face. Antibiotics are not routinely used.

Follow-Up

Pathology results must be transmitted to the patient. In tumors, it is important to maintain adequate follow-up for both recurrence and development of new tumors. This is often done by a dermatologist or a primary care physician.

Chapter 74

Removal of Subcutaneous Small Tumors, Cysts and Foreign Bodies

Neil A. Fine

Indications

Indications include lesions that are growing in size or painful, or are suspicious for malignancy due to their hard size, irregularity, or fixation to the overlying skin. Patients may desire removal of lesions for cosmetic reasons, but should understand that a scar is inevitable.

Preop

Discuss with the patient the nature of the scar and the unpredictability of scars. Patients who take aspirin or anticoagulants are advised to stop them. Often however, if the lesions are small this is not necessary. This procedure would commonly be performed in an office setting or in an ambulatory care setting using local anesthesia or as a conjunctive procedure as part of a more involved procedure requiring some other form of anesthesia.

Procedure

Step 1. Position the patient so the cyst or lesion is visible.

Step 2. Cleanse the skin with alcohol.

Step 3. Mark the overlying skin in the form of an ellipse if the lesion is protuberant. This will allow for excision of excess skin due to the stretching of the protuberant cystic lesion.

Step 4. Inject with lidocaine with epinephrine. Allow at least 5-10 minutes to pass prior to incision of the skin. Prep with betadine and sterile towels. Incise the skin with a 15 blade and switch to scissors or continue with the 15 blade down to the surface of the cyst or foreign body. Once at the level of the cyst or the foreign body, it is most advantageous to spread and dissect with scissors. Next, when the cyst or foreign body is freed circumferentially, try to remove in one piece. If it is not possible to remove in one piece, special care should be taken in the case of a cyst to ensure that all fragments of the true cyst itself are removed to decrease recurrence. Small bleeding points can be tied or sutured with small absorbable sutures. Alternatively, handheld cautery units may be utilized. Often due to the vascular constrictive effects of epinephrine, neither of these is necessary. The skin is closed in layers with deep buried suture followed by an external skin suture if necessary and then skin tapes.

Northwestern Handbook of Surgical Procedures, edited by Richard H. Bell, Jr. and Dixon B. Kaufman. ©2005 Landes Bioscience.

Comments

The patient should be warned that scars on the upper back, chest, and shoulders are particularly prone to forming hypertrophic scars or keloids. Any suspicious or unusual lesions need to be sent to pathology.

Postop

Pressure is held overlying the area for 15-20 minutes to decrease bruising. Patients are allowed to shower the next day.

Complications

Bleeding and infection are always possible with any excision, but they are quite rare with simple skin excisions. They are especially rare in the face. Antibiotics are not routinely used.

Follow-Up

Pathology results must be transmitted to the patient. In tumors, it is important to maintain adequate follow-up for both recurrence and development of new tumors. This is often done by a dermatologist or a primary care physician.

74

Section 5: Cardiothoracic Surgery

Section Editor: David Fullerton

Chapter 75

Esophagectomy: Ivor-Lewis

Sudhir Sundaresan

Indications

1. Esophageal malignancies involving the esophagogastric junction or mid-lower esophagus (at or below the azygos vein level).
2. Barrett's esophagus with high-grade dysplasia.
3. End-stage benign esophageal diseases (typically after multiple failed operations for gastroesophageal reflux or motor disorders).

Preop

1. Accurate diagnosis of the esophageal disease is accomplished by a detailed history, barium esophagogram, and endoscopy with biopsy.
2. In resections for esophageal cancer, rigorous tumor staging is necessary to determine that the lesion is localized and therefore amenable to a curative resection. This is accomplished by clinical evaluation (history and physical examination) and by computed tomographic (CT) scan of the chest and abdomen. Newer staging modalities include endoscopic ultrasound (to assess the degree of penetration of the esophageal wall by the tumor and the status of the locoregional lymph nodes), and positron emission tomography (PET) to identify distant bloodborne metastases. Preoperative surgical staging using thoracoscopy and laparoscopy has also been described but has not been universally accepted.
3. Fiberoptic and rigid bronchoscopy are necessary to rule out the possibility of airway involvement by the tumor.
4. Standard preoperative hematological and biochemical blood work and coagulation profile; urinalysis; electrocardiogram (EKG); type and screen. If the patient has documented cardiac disease or if the EKG is abnormal, a formal cardiology assessment should be conducted. If the patient has documented pulmonary disease, a significant smoking history, or if there is any significant breathing impairment, formal pulmonary function tests (PFT) and arterial blood gas (ABG) should be obtained.
5. Many centers advocate the use of preoperative (neoadjuvant) chemo-radiotherapy in esophageal carcinoma, although there are insufficient data to support this approach as the standard of care in this disease.

Procedure

Step 1. The first phase of the operation is performed with the patient positioned supine with arms tucked at the sides. Monitoring includes intravenous access, radial arterial line, Foley catheter, and central venous line inserted in the right side of the neck.

Step 2. The abdomen is prepared and draped as a large sterile field.

Northwestern Handbook of Surgical Procedures, edited by Richard H. Bell, Jr. and Dixon B. Kaufman. ©2005 Landes Bioscience.

75

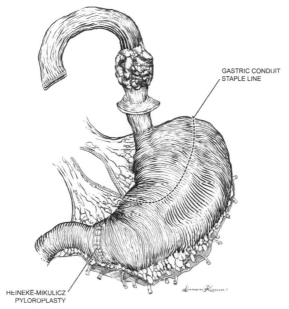

GASTRIC CONDUIT
STAPLE LINE

HEINEKE-MIKULICZ
PYLOROPLASTY

Figure 75.1. Ivor-Lewis esophagectomy. Abdominal part of the operation.

Step 3. An upper midline incision from the xiphoid to umbilicus is made. The xiphisternum is excised. The falciform ligament is divided, and a table-mounted retractor is inserted to exert strong upward and lateral pull on the costal arches bilaterally. A Balfour retractor is also inserted to improve the abdominal exposure.

Step 4. In resections for carcinoma, a full inspection of the abdomen and pelvis is performed to rule out the possibility of metastatic disease. Careful attention is paid to the liver, the primary tumor, and the upper abdominal lymph nodes. Once this has proven to be satisfactory, the resection and gastric pull-up are performed.

Step 5. The stomach is fully mobilized, in anticipation of being used as the replacement conduit for swallowing. The stomach will ultimately be perfused by the right gastroepiploic and right gastric arterial systems so extreme care must be taken throughout the dissection not to damage these vessels. This mobilization begins with division of the lesser omentum (with cautery) starting at the mid-lesser curvature of the stomach and ending at the right crus of the diaphragm. The distal esophagus is dissected out and encircled with a Penrose drain.

Step 6. The greater curvature is fully mobilized. The gastrocolic ligament is divided with cautery to enter the lesser sac, staying a generous distance away from the gastroepiploic arcade. The posterior stomach wall and the transverse mesocolon must be accurately visualized. The greater curve can be mobilized with cautery, with the harmonic scalpel being reserved for larger vessels (e.g., short gastric vessels). The mobilization is taken to the right just beyond the pylorus. The duodenum is fully Kocherized using cautery, giving enough mobility to allow the pylorus to lie near the esophageal hiatus after the stomach is transposed to the neck. All of the short gastric vessels are divided, to arrive at the left crus of the diaphragm. The large

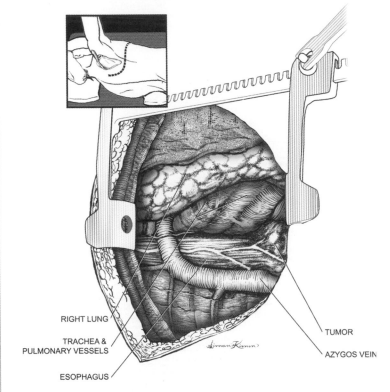

RIGHT LUNG

TRACHEA &
PULMONARY VESSELS

ESOPHAGUS

TUMOR

AZYGOS VEIN

Figure 75.2. Ivor-Lewis esophagectomy. Thoracic part of the operation.

posterior vascular pedicle containing the left gastric vessels and lymph nodes is isolated at the upper border of the pancreas, thinned into two or three smaller stalks, and divided using sequential applications of the linear vascular stapler.

Step 7. Pyloroplasty is performed using the Heinecke-Mikulicz technique. Stay sutures of 3-0 silk are placed on either side of the pylorus. The pylorus is then opened longitudinally using cautery, slightly onto the duodenal and gastric sides, and ensuring that all of the pyloric muscle has been ablated. The opening is then closed transversely, using a series of full-thickness sutures of 2-0 silk, taking care to include bites of the mucosa.

Step 8. The esophageal hiatus is enlarged. This is done using cautery after dividing the phrenic vein which is routinely identified crossing anterior to the hiatus. The esophagus is then freed up well above the esophageal hiatus. The operating table is tipped into Trendelenburg position to improve the exposure up into the mediastinum. The esophagus is then freed up using cautery or the harmonic scalpel.

Step 9. A feeding jejunostomy catheter is placed, using a standard Witzel technique and a 12 F red rubber catheter. The ligament of Treitz is identified on the inferior surface of the transverse mesocolon, and the catheter is typically inserted about one foot distal to this point. A pursestring suture of 3-0 silk is used in the antimesenteric border of the bowel to secure the catheter at the entry site. A series of interrupted 3-0 silk Lembert sutures are used to bury the catheter in the bowel wall over a short distance and to secure the bowel to the parietal peritoneum. The catheter is brought out through a small stab wound in the left midabdomen and secured at the skin with a 3-0 nylon suture.

Step 10. The laparotomy incision is closed with continuous #2 polypropylene suture in the fascia and staples in the skin. Sterile dressings are applied and the drapes are taken down.

Step 11. The right thoracotomy is performed. A left-sided double-lumen tube, or alternately a right-sided bronchial blocker, must be placed by the anesthesiologist between the abdominal and thoracic phases of the operation. After this is complete, the patient is repositioned in the left lateral decubitus position. The right chest is prepared and draped as a sterile field. A right posterolateral thoracotomy incision is made, entering the chest in the 5th (or preferably 4th) intercostal space. The serratus anterior muscle can be spared during creation of the thoracotomy. Retractors are inserted at right angles to one another to provide the exposure. The right lung is deflated and packed anteriorly with moist laparotomy sponges.

Step 12. The thoracic esophagus is mobilized, the esophagectomy is completed, and the gastric tube is prepared. The azygos vein arch is dissected out and the vein is divided between heavy silk ligatures. The esophagus is dissected out and encircled with a large Penrose drain. The esophagus is then fully mobilized in the chest to the desired level of transection [usually near the thoracic inlet]. Below the carina, the segmental blood supply to the esophagus is isolated as a series of small stalks and divided distal to clips or silk ligatures. This method is more secure than dividing the stalks with cautery alone, as an unrecognized lymphatic injury can result in postoperative chylothorax. When mobilizing the esophagus above the carina, it is vital to keep the dissection immediately on the adventitia of the esophagus to minimize the chance of operative injury to the left recurrent laryngeal nerve. Traction is applied to the distal esophagus to deliver the stomach into the chest, taking care to maintain the proper orientation (lesser curvature facing towards the patient's right side). The esophagus is transected proximally with a linear stapler. The lesser curvature is then resected along with the esophagus, using sequential firings of the linear stapler. The esophagogastrectomy specimen is passed off the field. If resection is for tumor, a frozen section is obtained to verify that the proximal margin is clear of tumor. The staple line along the lesser curve is then inverted using a continuous Lembert suture of 4-0 polypropylene.

Step 13. The esophagogastrostomy is created, usually midway between the carina and the thoracic inlet. It is desirable to create the anastomosis as high as reasonably possible since this will decrease the chance of troublesome postoperative reflux and associated complications (e.g., peptic esophagitis or stricture). The anastomosis begins with excision of the staple line from the esophagus and creation of a small gastrotomy on the upper greater curve (staying away from the lesser curve suture line). A common method of hand-sewn anastomosis involves two layers of simple interrupted 4-0 silk sutures. Sutures of the inner layer incorporate the esophageal mucosa to the full thickness of the stomach and are placed so that knots are tied on the inside to maximize mucosal inversion. The outer row of sutures approximates the esophageal muscle to the seromuscular layer of the stomach and buries the inner suture line. An alternate technique involves the use of the linear stapler to create a functional end-to-end anastomosis. The endo-GIA stapler, with a 45 mm long blue cartridge (3.5 mm staples) is inserted into both the esophagus and stomach, which have been laid in apposition. The stapler is fired, creating most of the anastomosis and leaving a common opening into the esophagus and stomach. This common opening is closed using a continuous 4-0 monofilament absorbable suture, with a second layer of 4-0 silk sutures (approximating the esophageal muscle to the seromuscular layer of the stomach) used to bury this suture line. In either technique, a nasogastric (NG) tube must be passed by the anesthesiologist, guided across the anastomosis, and then advanced into the lower portion of the gastric tube before completion of the anastomosis.

Step 14. Two 28 F chest tubes are placed, brought out through separate stab wounds inferior to the main incision, secured to the skin with heavy polypropylene suture, and connected to a drainage apparatus. One tube lies anterior to the lung. The other is positioned posteriorly and lies adjacent to the anastomosis. It is secured here with several 3-0 chromic catgut sutures to the pleura to prevent it from moving. This tube is intended to provide external drainage in case an anastomotic leak develops later.

Step 15. The right lung is manually reinflated and the thoracotomy is closed, using #2 Vicryl sutures as pericostal sutures, #1 Vicryl in the extracostal fascial layers, 2-0 Vicryl in subcutaneous tissues, and staples in the skin. Absorbent gauze dressings are applied and the drapes are taken down.

Postop

1. *Intravenous fluids.* Generous volume resuscitation is required in the first 24 hours postoperatively (e.g., normal saline at 150-175 ml/h) since the anticipated 3rd space volume loss will be substantial. Urine output is carefully monitored, with 0.5-1.0 ml/kg/h considered adequate. ABGs should be obtained periodically to monitor PaO_2 and $PaCO_2$ (during ventilator weaning; see below), and to monitor the base deficit, another useful parameter to assess the adequacy of the circulation. Maintenance of satisfactory circulation is vital, since the upper portion of the gastric conduit is perfused mostly by intramural blood supply. This is best achieved by generous use of crystalloids (as described above), and transfusion of packed red cells (as dictated by low hemoglobin values). About 24-36 hours after surgery, the 3rd space volume loss subsides, and the extravascular volume is returned to the intravascular space. This is heralded by diuresis and signals the appropriate timing to reduce volume administration to maintenance fluids and judicious use of diuretics (e.g., furosemide 20 mg IV Q 8-12 h). Diuretics are continued until the patient has reached their preoperative weight.

2. *Respiratory status.* The patient is maintained on a ventilator postoperatively and weaned and extubated the following morning. Supplemental oxygen is then given by face mask and ultimately by nasal prongs until it is stopped. ABGs are checked during the phase of ventilator weaning and immediately after extubation. If the patient is stable thereafter, pulse oximetry is sufficient to guide the administration of oxygen. During the 48 hours following extubation, nebulized bronchodilators and chest physical therapy are used preemptively to minimize the development of serious sputum retention and pneumonia. Ambulation is begun on the morning of postoperative day 2. Deep vein thrombosis (DVT) prophylaxis (consisting of heparin 5000 U sc BID and sequential compression devices [SCDs]) are started immediately postoperatively. Once ambulation is

well established, the SCDs can be removed and subcutaneous heparin is continued until discharge from the hospital. Daily chest X-rays are obtained for several days to monitor for development of significant atelectasis, consolidation, pleural effusion, and position of the gastric conduit.

3. *Pain management.* A thoracic epidural catheter is inserted preoperatively and maintained for 3 days for optimal pain relief. Subsequently, a patient-controlled analgesia (PCA) pump is utilized. After oral intake is resumed, elixir can be taken by mouth and is prescribed for home use after hospital discharge.

4. *GI tract management.* The patient is kept NPO with a sumping nasogastric (NG) tube on continuous suction until postoperative day 5. If the clinical progress is satisfactory, then a contrast swallow radiological evaluation is performed to assess anastomotic integrity. Water-soluble contrast material is swallowed first, and if X-rays are negative for leak, then they are repeated as thin barium is swallowed. If this sequence of X-rays is negative for anastomotic leak, the NG tube is removed and oral intake of water is started on postoperative day 5. The subsequent advancement of oral intake is dependent on the patient's clinical progress, but the general plan is as follows: postoperative day 6, clear liquids; postoperative day 7, full liquids; postoperative day 8, mechanical soft diet. Five percent dextrose and water (D5W) is started through the jejunostomy catheter on postoperative day 2, and tube feeds are started on postoperative day 3 and gradually advanced to the target rate over the ensuing 36 hours. If the patient's oral intake of soft food is sufficient by postoperative day 9, they may be discharged from the hospital without home jejunostomy feeds. They are instructed to simply flush the catheter with tap water BID until their return visit to the outpatient clinic at which time the catheter is removed. If the patient's oral intake of protein and calories is insufficient, the jejunostomy catheter may be used for home tube feeds for a few days/weeks, until oral intake has improved.

5. *Antibiotics.* Clindamycin (600 mg IV Q 6 h) is given prophylactically for 48 hours then stopped.

6. *Wound care.* Skin staples are removed on the day of discharge. Chest tubes are initially kept on 20 cm water suction. The anterior chest tube may be removed by postoperative day 2. The posterior chest tube draining the region of the anastomosis is removed after the patient's oral intake has progressed to soft solids.

Complications

1. Medical complications following major surgery include: atelectasis, pneumonia, deep vein thrombosis (DVT) and pulmonary embolism (PE), atrial fibrillation, myocardial infarction, and stroke.

2. General surgical complications of such major surgery include bleeding (requiring transfusion) and wound infection(s).

3. Complications pertinent to Ivor-Lewis esophagectomy include ischemic necrosis of the gastric conduit, with or without mediastinitis or empyema; anastomotic leak, with or without mediastinitis or empyema; left recurrent laryngeal nerve injury; injury to thoracic duct with ensuing chylothorax; splenic injury necessitating splenectomy; and pleural injury causing left pneumothorax and/or pleural effusion.

4. Since the thoracic esophagus is dissected free under direct vision using a right thoracotomy, the risk of intraoperative injury to the membranous airway or major vessels (e.g., aorta) should be very low.

Follow-Up

75

1. Postoperative follow-up at 2-3 weeks entails clinical examination, chest X-ray, and removal of jejunostomy catheter.
2. If postoperative adjuvant therapy (chemotherapy, radiation therapy) is recommended, it can be started at about 4 weeks postoperatively.
3. Subsequent follow-up is at 3-monthly intervals that includes clinical examination and chest X-ray. If esophagectomy was performed for carcinoma, follow-up must be conducted for a total of 5 years after conclusion of treatment. CT scans may be obtained to monitor for recurrence at 6-12 month intervals, assuming that clinical progress is otherwise favorable.

Chapter 76

Esophagectomy: Left Transthoracic

Sudhir Sundaresan

Indications

Left transthoracic esophagectomy is indicated for resection of esophageal malignancies involving the distal esophagus or esophagogastric junction, particularly when the tumor involves a substantial part of the proximal stomach, thus limiting the amount of stomach available to pull up.

Preop

1. Accurate diagnosis of the esophageal disease is accomplished by a detailed history, barium esophagogram, and endoscopy with biopsy.
2. In resection for esophageal cancer, rigorous tumor staging is necessary to determine that the lesion is localized and therefore amenable to a curative resection. This is accomplished by clinical evaluation (history and physical examination) and by computed tomographic (CT) scan of the chest and abdomen. Newer staging modalities include endoscopic ultrasound (to assess the degree of penetration of the esophageal wall by the tumor and the status of the locoregional lymph nodes), and positron emission tomography (PET scanning) to identify distant bloodborne metastases. Preoperative surgical staging using thoracoscopy and laparoscopy has also been described but has not been universally accepted.
3. Standard preoperative hematological and biochemical blood work and coagulation profile; urinalysis; electrocardiogram (EKG); type and screen. If the patient has documented cardiac disease or if the EKG is abnormal, a formal cardiology assessment should be conducted. If the patient has documented pulmonary disease, a significant smoking history, or if there is any significant breathing impairment, formal pulmonary function tests (PFT) and arterial blood gas (ABG) should be obtained.
4. Many centers advocate the use of preoperative (neoadjuvant) chemo-radiotherapy in esophageal carcinoma, although there are insufficient data to support this approach as the standard of care in this disease.

Procedure

Step 1. The operation can be performed in one of three ways. The first option entails separate midline laparotomy and left thoracotomy approaches. This obviously requires repositioning of the patient and creation of a separate sterile field for the left thoracic portion. In this scenario, the operation is virtually identical to the Ivor-Lewis resection and begins with Steps 1-11 already described for that procedure, except that a left thoracotomy is utilized, entering the chest typically in the 6th intercostal space. The subsequent conduct of the thoracic portion is also similar to Steps 12 and 13 for the Ivor-Lewis resection, except that the esophagus is

76

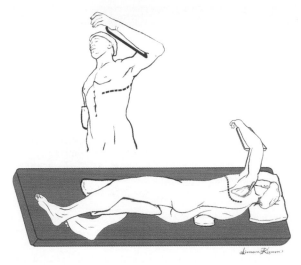

Figure 76.1. Left transthoracic esophagectomy through a left posterolateral thoracotomy.

mobilized through a left thoracotomy to just below the arch of the aorta. The esophagus must be carefully freed from the pericardium, inferior pulmonary vein, and left main bronchus, and attention is paid to carefully securing the segmental blood supply as already described. The left vagus nerve must be separated from the esophagus below the origin of the recurrent branch in order to spare it. The anastomotic technique, chest tube use, and thoracotomy closure are identical to that described for the Ivor-Lewis resection (Steps 13-15).

Step 2. The second option entails a left posterolateral thoracotomy alone, entering the chest in the 6th intercostal space. The periphery of the left diaphragm is then opened as follows: Paired stay sutures of 3-0 polypropylene are placed on either side of the proposed phrenotomy. Cautery is used to make a full-thickness opening, conforming to the curvature of the periphery of the diaphragm and leaving a small rim attached to the chest wall to facilitate its subsequent reconstitution. The stay sutures are useful to retract the diaphragm. The exposure of the upper abdomen through this technique is superb and will facilitate mobilization of the entire stomach (Ivor-Lewis resection, Steps 5-9) except for the Kocher maneuver and pyloroplasty. Since the gastric tube is required to reach only as high as the aortic arch, the elimination of the Kocher maneuver here is acceptable. The delivery of the stomach via the hiatus into the chest, completion of the resection, preparation of the gastric tube, creation of the anastomosis, chest tube placement, and closure are as described previously. The only additional step is the closure of the phrenotomy. This is done using a series of horizontal mattress sutures of #1 polypropylene, followed by a continuous #1 polypropylene suture line ("vest-over-pants" closure).

Step 3. The third option entails a left thoracoabdominal incision. This approach is less favored since the incision involves division of the costal arch, which is associated with considerable late morbidity (pain). The main advantage of this approach

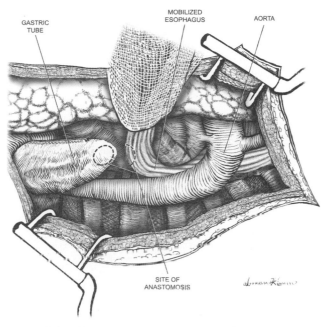

Figure 76.2. Left transthoracic esophagectomy. Site of gastroesophageal anastomosis.

(and the reason that many surgeons use this approach preferentially for total esophagectomy with neck anastomosis) is that it improves the abdominal exposure, particularly to the right of the midline, and will therefore facilitate a complete Kocher maneuver and pyloroplasty. The remaining steps described below will detail this approach.

Step 4. The patient is positioned in full right lateral decubitus position as for left thoracotomy. A sterile field is created that includes the chest and abdomen slightly to the right of the midline and below the umbilicus.

Step 5. With the left lung deflated, a 6th or 7th intercostal space thoracotomy is created and taken obliquely across the costal arch, ending near the midline of the epigastrium, midway between the xiphisternum and the umbilicus. The costal arch is divided sharply with a scalpel. The abdominal obliques and the anterior and posterior layers of the left rectus sheath are also divided, although the rectus abdominis muscle is simply retracted to the right. A small circumferential opening in the periphery of the left diaphragm is also necessary to provide exposure (see Step 2). A chest retractor (for the thoracotomy) and a Balfour retractor (for the abdomen) will provide superb exposure. The process of gastric mobilization, pyloroplasty, Kocher maneuver, jejunostomy, esophagogastrectomy, preparation and transposition of the gastric tube, and esophagogastric anastomosis have already been described.

Step 6. Once the operation is complete, the chest tube insertion and wound closure is performed as already described. The abdomen is closed with #1 polypropylene

76

in the anterior and posterior layers of the rectus sheath, in the obliques, and in the diaphragm (diaphragm closure described above). A short plug of cartilage is excised at the costal arch to eliminate any overriding of the divided ends, and a figure-of-eight suture of heavy polypropylene is used to secure the arch at this point, in addition to the remaining pericostal sutures. The technique of chest closure has already been described. The left lung is manually reinflated prior to completion of closure. Absorbent gauze dressings are applied to the incision and the drapes are taken down.

Postop

1. *Intravenous fluids.* Generous volume resuscitation is required in the first 24 hours postoperatively (e.g., normal saline at 150-175 ml/h) since the anticipated 3rd space volume loss will be substantial. Urine output is carefully monitored, with 0.5-1.0 ml/kg/h considered adequate. ABGs should be obtained periodically to monitor PaO_2 and $PaCO_2$ (during ventilator weaning; see below), and to monitor the base deficit, another useful parameter to assess the adequacy of the circulation. Maintenance of satisfactory circulation is vital, since the upper portion of the gastric conduit is perfused mostly by intramural blood supply. This is best achieved by generous use of crystalloids (as described above), and transfusion of packed red cells (as dictated by low hemoglobin values). About 24-36 hours after surgery, the 3rd space volume loss subsides, and the extravascular volume is returned to the intravascular space. This is heralded by diuresis, and signals the appropriate time to reduce volume administration to maintenance fluids and judicious use of diuretics (e.g., furosemide 20 mg IV Q 8-12 h). Diuretics are continued until the patient has reached their preoperative weight.

2. *Respiratory status.* The patient is maintained on a ventilator postoperatively and weaned and extubated the following morning. Supplemental oxygen is then given by face mask and ultimately by nasal prongs until it is stopped. ABGs are checked during the phase of ventilator weaning and immediately after extubation. If the patient is stable thereafter, pulse oximetry is sufficient to guide the administration of oxygen. During the 48 hours following extubation, nebulized bronchodilators and chest physical therapy are used preemptively to minimize the development of serious sputum retention and pneumonia. Ambulation is begun on the morning of the 2nd postoperative day. Deep vein thrombosis (DVT) prophylaxis (consisting of heparin 5000 U sc BID, and sequential compression devices [SCDs]) are started immediately postoperatively. Once ambulation is well established, the SCDs can be removed and subcutaneous heparin is continued until discharge from the hospital. Daily chest X-rays are obtained for several days to monitor for development of significant atelectasis, consolidation, pleural effusion, and position of the gastric conduit.

3. *Pain management.* A thoracic epidural catheter is inserted preoperatively and maintained for 3 days for optimal pain relief. Subsequently, a patient-controlled analgesia (PCA) pump is utilized. After oral intake is resumed, elixir can be taken by mouth and is prescribed for home use after hospital discharge.

4. *GI tract management.* The patient is kept NPO with a sumping nasogastric (NG) tube on continuous suction until postoperative day 5. If the clinical progress is satisfactory, a contrast swallow radiological evaluation is performed to assess anastomotic integrity. Water-soluble contrast material is swallowed first, and if X-rays are negative for leak, they are repeated as thin barium is swallowed. If

this sequence of X-rays is negative for anastomotic leak, the NG tube is removed and oral intake of water is started on postoperative day 5. The subsequent advancement of oral intake is dependent on the patient's clinical progress, but the general plan is as follows: postoperative day 6, clear liquids; postoperative day 7, full liquids; postoperative day 8, mechanical soft diet. Five percent dextrose and water (D5W) is started through the jejunostomy catheter on postoperative day 2, and tube feeds are started on postoperative day 3 and gradually advanced to the target rate over the ensuing 36 hours. If the patient's oral intake of soft food is sufficient by postoperative day 9, she may be discharged from the hospital without home jejunostomy feeds. Patients are instructed to simply flush the catheter with tap water BID until their return visit to the outpatient clinic at which time the catheter is removed. If the patient's oral intake of protein and calories is insufficient, then the jejunostomy catheter may be used for home tube feeds for a few days/weeks until oral intake has improved.

5. *Antibiotics.* Clindamycin (600 mg IV Q 6 h) is given prophylactically for 48 hours then stopped.

6. *Wound care.* Skin staples are removed on the day of discharge. Chest tubes are initially kept on 20 cm water suction. The anterior chest tube may be removed by postoperative day 2. The posterior chest tube (draining the region of the anastomosis) is removed after the patient's oral intake has progressed to soft solids.

Complications

1. Medical complications following major surgery include: atelectasis, pneumonia, deep vein thrombosis (DVT) and pulmonary embolism (PE), atrial fibrillation, myocardial infarction, and stroke.

2. General surgical complications of such major surgery include bleeding (requiring transfusion) and wound infection(s).

3. Complications pertinent to left transthoracic esophagectomy include ischemic necrosis of the gastric conduit, with or without mediastinitis or empyema; anastomotic leak, with or without mediastinitis or empyema; left recurrent laryngeal nerve injury; injury to thoracic duct with ensuing chylothorax; splenic injury necessitating splenectomy; and pleural injury causing right pneumothorax and/or pleural effusion.

4. Since the thoracic esophagus is dissected free under direct vision using a left thoracotomy, the risk of intraoperative injury to the membranous airway or major vessels (e.g., aorta) should be very low.

Follow-Up

1. Postoperative follow-up at 2-3 weeks entails clinical examination, chest X-ray, and removal of jejunostomy catheter.

2. If postoperative adjuvant therapy (chemotherapy, radiation therapy) is recommended, it can be started at about 4 weeks postoperatively.

3. Subsequent follow-up is at 3-month intervals and includes clinical examination and chest X-ray. If esophagectomy was performed for carcinoma, follow-up must be conducted for a total of 5 years after conclusion of treatment. CT scans may be obtained to monitor for recurrence at 6-12 month intervals, assuming that clinical progress is otherwise favorable.

Chapter 77

Esophagectomy: Transhiatal

Sudhir Sundaresan

Indications

1. Esophageal malignancies involving the esophagogastric junction, lower third of the esophagus at or below the inferior pulmonary vein level, or upper esophagus at or above the thoracic inlet. In the latter situation, the tumor resection usually requires pharyngolaryngoesophagectomy in order to achieve a satisfactory proximal margin, and incorporates a transhiatal esophagectomy with gastric pull-up to the hypopharynx.
2. Barrett's esophagus with high-grade dysplasia.
3. End-stage benign esophageal diseases (typically after multiple failed operations for gastroesophageal reflux or motor disorders). Transhiatal esophagectomy is contraindicated for the resection of midesophageal tumors lying adjacent to the membranous trachea or left main bronchus.

Preop

1. Accurate diagnosis of the esophageal disease is accomplished by a detailed history, barium esophagogram, and endoscopy with biopsy.
2. In resections for esophageal cancer, rigorous tumor staging is necessary to determine that the lesion is localized and therefore amenable to a curative resection. This is accomplished by clinical evaluation (history and physical examination) and by computed tomographic (CT) scan of the chest and abdomen. Newer staging modalities include endoscopic ultrasound to assess the degree of penetration of the esophageal wall by the tumor and the status of the locoregional lymph nodes and positron emission tomography (PET scanning) to identify distant bloodborne metastases. Preoperative surgical staging using thoracoscopy and laparoscopy has also been described but has not been universally accepted.
3. Fiberoptic and rigid bronchoscopy are necessary to rule out the possibility of airway involvement by the tumor.
4. Standard preoperative hematological and biochemical blood work and coagulation profile; urinalysis; electrocardiogram (EKG); type and screen. If the patient has documented cardiac disease or if the EKG is abnormal, a formal cardiology assessment should be conducted. If the patient has documented pulmonary disease, a significant smoking history, or if there is any significant breathing impairment, formal pulmonary function tests (PFT) and arterial blood gas (ABG) should be obtained.
5. Many centers advocate the use of preoperative neoadjuvant chemo-radiotherapy in esophageal carcinoma. There are insufficient data to support this approach as the standard of care in this disease.

Northwestern Handbook of Surgical Procedures, edited by Richard H. Bell, Jr. and Dixon B. Kaufman. ©2005 Landes Bioscience.

Procedure

Step 1. The patient is positioned supine with arms tucked at the sides. An inflatable mattress is placed beneath the shoulders and inflated to achieve neck extension, with the head turned slightly to the right. Monitoring includes intravenous access, radial arterial line, Foley catheter, and central venous line inserted in the right side of the neck.

Step 2. The neck, chest, and abdomen are prepared and draped as one large sterile field.

Step 3. An upper midline laparotomy is made from the xiphoid to the umbilicus. The xiphisternum is excised. The falciform ligament is divided, and a table-mounted retractor is inserted to exert strong upward and lateral pull on the costal arches bilaterally. A Balfour retractor is also inserted to improve the abdominal exposure.

Step 4. In resections for carcinoma, full inspection of the abdomen and pelvis is performed to rule out the possibility of metastatic disease. Careful attention is paid to the liver, the primary tumor, and the upper abdominal lymph nodes. Once this has proven to be satisfactory, the resection and gastric pull-up are performed.

Step 5. The stomach is fully mobilized, in anticipation of being used as the replacement conduit for swallowing. The stomach will ultimately be perfused by the right gastroepiploic and right gastric arterial systems, so extreme care must be taken throughout the dissection not to damage these vessels. This mobilization begins with division of the lesser omentum (with cautery) starting at the mid-lesser curvature of the stomach and ending at the right crus of the diaphragm. The distal esophagus is dissected out and encircled with a Penrose drain.

Step 6. The greater curvature is fully mobilized. The gastrocolic ligament is divided with cautery to enter the lesser sac, staying a generous distance away from the gastroepiploic arcade. The posterior stomach wall and the transverse mesocolon must be accurately visualized. The greater curve can be mobilized with cautery, with the harmonic scalpel being reserved for larger vessels (e.g., short gastric vessels). Mobilization is taken to the right just beyond the pylorus. The duodenum is fully kocherized using cautery, giving enough mobility to allow the pylorus to lie near the esophageal hiatus after the stomach is transposed to the neck. All of the short gastric vessels are divided to arrive at the left crus of the diaphragm. The large posterior vascular pedicle containing the left gastric vessels and lymph nodes is isolated at the upper border of the pancreas, thinned into two or three smaller stalks, and divided using sequential applications of the linear vascular stapler. The stomach is now fully mobile, and it must be verified that it will reach the neck without limitation.

Step 7. Pyloroplasty is performed using the Heinecke-Mikulicz technique. Stay sutures of 3-0 silk are placed on either side of the pylorus. The pylorus is then opened longitudinally using cautery, slightly onto the duodenal and gastric sides, and ensuring that all of the pyloric muscle has been ablated. The opening is then closed transversely, using a series of full-thickness sutures of 2-0 silk, taking care to include bites of the mucosa.

Step 8. The left neck is opened to permit mobilization of the cervical esophagus. An oblique incision is made along the anterior border of the left sternocleidomastoid (SCM) muscle and deepened with cautery through the platysma. The incision starts at the level of the brow of the thyroid cartilage and includes a short hockey-stick extension into the suprasternal notch. The SCM is mobilized laterally to expose the omohyoid muscle which is divided. The fascia along the lateral border of the strap muscles is divided, and the strap muscles are mobilized and divided with cautery near their sternal attachments. Gentle digital pressure is applied to retract the trachea to the right. The middle thyroid vein and inferior thyroid artery are identified and divided between 3-0 silk ligatures. Dissection is then carried out on the adventitia of the esophagus to isolate it and encircle it with a Penrose drain. Care is taken to identify the left recurrent laryngeal nerve in the tracheoesophageal groove and preserve it; there must be no cautery or pressure from metal instruments applied to the nerve.

Step 9. The mediastinal portion of the esophagus is fully freed up. Working from the neck, blunt dissection (using the finger tip or a sponge forcep) is used to free the esophagus to the level of the carina. The success of this mobilization depends on accurately isolating the esophagus in the correct plane (Step 8). This maneuver is safe since there is no segmental esophageal blood supply at this level. Keeping the blunt dissection immediately on the esophagus minimizes the chance of injuries to the recurrent nerve, membranous trachea, or adjacent blood vessels. From the abdominal aspect, the esophageal hiatus must be enlarged. This is done using cautery after dividing the phrenic vein which is routinely identified crossing anterior to the hiatus. The operating table is tipped into Trendelenburg position to improve the exposure up into the mediastinum. The esophagus is then freed up using cautery or the harmonic scalpel. The last portion of mobilization at the carinal level may require blunt digital dissection.

Step 10. The esophagectomy is completed and the gastric tube is prepared. Strong traction is applied to the cervical esophagus to deliver the esophagus maximally out of the neck incision. The esophagus is transected as distal as possible with a linear stapler, with a long length of umbilical tape fastened to the distal stump. The distal esophagus is then delivered into the abdomen, leaving the umbilical tape traversing the mediastinum along the future path of transposition of the gastric tube. The lesser curvature is then resected along with the esophagus, using sequential firings of the linear stapler. The esophagogastrectomy specimen is passed off the field. If resection is for tumor, a frozen section is obtained to verify that the proximal margin is clear of tumor. The staple line along the lesser curvature is then inverted using a continuous Lembert suture of 4-0 polypropylene.

Step 11. The gastric tube is transposed to the neck. A laparoscopy camera bag is snugged around the stomach (like an accordion) after first drying the serosa of the stomach with a lap sponge. Drying the serosa is important since the plastic camera bag will drag the gastric tube to the neck by virtue of the friction it exerts on the stomach. The umbilical tape is tied to the upper end of the plastic bag. Constant firm tension is applied to the tape from the neck. The tip of the bag (containing the leading edge of the stomach tube) is gently guided into the esophageal hiatus, and the outside of the bag is lubricated by irrigating it with saline. As the bag emerges from the neck incision, the tip of the gastric fundus will be seen. At this point the bag can be cut away, pulled through, and discarded. The gastric tip is securely grasped with a Babcock clamp in preparation for the anastomosis. Care is taken at this juncture to verify that there has been no rotation of the stomach during the transposition maneuver; the greater curvature should be oriented to the patient's left, and the lesser curvature suture line should be facing the right, both in the neck and in the abdomen. The pyloroplasty should be just below the esophageal hiatus.

Step 12. The cervical esophagogastrostomy is created. It begins with excision of the staple line from the esophagus and creation of a small gastrotomy on the upper greater curve (staying away from the lesser curve suture line). A common method of hand-sewn anastomosis involves two layers of simple interrupted 4-0 silk sutures. Sutures of the inner layer incorporate the esophageal mucosa to the full thickness of the stomach and are placed so that knots are tied on the inside, to maximize mucosal inversion. The outer row of sutures approximates the esophageal muscle to the seromuscular layer of the stomach and buries the inner suture line. An alternate technique involves the use of the linear stapler to create a functional end-to-end anastomosis. The endo-GIA stapler, with a 45 mm long blue cartridge (3.5 mm staples) is inserted into both the esophagus and stomach, which have been laid in apposition. The stapler is fired, creating most of the anastomosis and leaving a common opening into the esophagus and stomach. This common opening is closed using a continuous 4-0 monofilament absorbable suture, with a second layer of 4-0 silk sutures (approximating the esophageal muscle to the seromuscular layer of the stomach) used to bury this suture line. In either technique, a nasogastric (NG) tube must be passed by the anesthesiologist, guided across the anastomosis, and then advanced into the lower portion of the gastric tube, before completion of the anastomosis. The anastomosis, along with the redundant esophagus and excess stomach, are all returned to the mediastinum below the thoracic inlet.

Step 13. A feeding jejunostomy catheter is placed, using a standard Witzel technique and a 12 F red rubber catheter. The ligament of Treitz is identified on the inferior surface of the transverse mesocolon, and the catheter is typically inserted about one foot distal to this point. A pursestring suture of 3-0 silk is used in the antimesenteric border of the bowel to secure the catheter at the entry site. A series of interrupted 3-0 silk Lembert sutures is used to bury the catheter in the bowel wall over a short distance and to secure the bowel to the parietal peritoneum. The catheter is brought out through a small stab wound in the left midabdomen and secured at the skin with a 3-0 nylon suture.

Step 14. The incisions are closed. A Penrose drain is guided down into the mediastinum to the anastomosis from the lower portion of the neck incision. The platysma layer is closed with continuous 2-0 Vicryl suture, deliberately leaving a generous defect at the Penrose site. This is done to facilitate establishment of a cutaneous fistula in the event of an anastomotic leak later. Skin is closed with staples. The laparotomy incision is closed with continuous #2 polypropylene suture in the fascia and staples in the skin. Absorbent gauze dressings are placed to both incisions.

Postop

1. *Intravenous fluids.* Generous volume resuscitation is required in the first 24 hours postoperatively (e.g., normal saline at 150-175 ml/h) since the anticipated 3rd space volume loss will be substantial. Urine output is carefully monitored, with 0.5-1.0 ml/kg/h considered adequate. ABGs should be obtained periodically to monitor PaO_2 and $PaCO_2$ (during ventilator weaning; see below) and to monitor the base deficit, another useful parameter to assess the adequacy of the circulation. Maintenance of satisfactory circulation is vital, since the upper portion of the gastric conduit is perfused mostly by intramural blood. This is best achieved by generous use of crystalloids (as described above) and transfusion of packed

red cells (as dictated by low hemoglobin values). Twenty-four to 36 hours after surgery, the 3rd space volume loss subsides, and the extravascular volume is returned to the intravascular space. This is heralded by diuresis and signals the appropriate timing to reduce volume administration to maintenance fluids and the judicious use of diuretics (e.g., furosemide 20 mg IV Q 8-12h). Diuretics are continued until the patient has reached their preoperative weight.

2. *Respiratory status.* The patient is maintained on a ventilator postoperatively and weaned and extubated the following morning. Supplemental oxygen is then given by face mask and ultimately by nasal prongs until it is stopped. ABGs are checked during the phase of ventilator weaning and immediately after extubation. If the patient is stable thereafter, pulse oximetry is sufficient to guide the administration of oxygen. For 48 hours following extubation, nebulized bronchodilators and chest physical therapy are used to minimize the development of serious sputum retention and pneumonia. Ambulation is begun on the morning of the second postoperative day. Deep vein thrombosis (DVT) prophylaxis (consisting of heparin 5000 U sc BID, and sequential compression devices [SCDs]) is started immediately postop. Once ambulation is well established, the SCDs can be removed and subcutaneous heparin is continued until discharge from the hospital. Daily chest X-rays are obtained for several days to monitor for development of significant atelectasis, consolidation, pleural effusion, and position of the gastric conduit.

3. *Pain management.* A thoracic epidural catheter is inserted preoperatively and maintained for 3 days for optimal pain relief. Subsequently, a patient-controlled analgesia (PCA) pump is utilized. After oral intake is resumed, elixir can be taken by mouth and is prescribed for home use after hospital discharge.

4. *GI tract management.* The patient is kept NPO with a sumping nasogastric (NG) tube on continuous suction until postoperative day 5. If the clinical progress is satisfactory, a contrast swallow radiological evaluation is performed to assess anastomotic integrity. Water-soluble contrast material is swallowed first, and if X-rays are negative for leak, then they are repeated with a thin barium swallow. If this sequence of X-rays is negative for anastomotic leak, the NG tube is removed and oral intake of water is started on postoperative day 5. The subsequent advancement of oral intake is dependent on the patient's clinical progress, but the general plan is as follows: postoperative day 6, clear liquids; postoperative day 7, full liquids; postoperative day 8, mechanical soft diet. Five percent dextrose and water (D5W) is started through the jejunostomy catheter on postoperative day 2, and tube feeds are started on postoperative day 3 and gradually advanced to the target rate over the ensuing 36 hours. If the patient's oral intake of soft food is sufficient by postoperative day 9, she may be discharged from the hospital without home jejunostomy feeds and is instructed to simply flush the catheter with tap water twice a day until the return visit to the outpatient clinic, at which time the catheter is removed. If the patient's oral intake of protein and calories is insufficient, the jejunostomy catheter may be used for home tube feeds for a few days/weeks until oral intake has improved.

5. *Antibiotics.* Clindamycin (600 mg IV Q 6h) is given prophylactically for 48 hours then stopped.

6. *Wound care.* The Penrose drain in the neck is gradually shortened starting on postoperative day 6 and is removed on postoperative day 9 (after several days of oral intake). Skin staples are removed on the day of discharge.

Complications

1. Medical complications following major surgery include: atelectasis, pneumonia, deep vein thrombosis (DVT) and pulmonary embolism (PE), atrial fibrillation, myocardial infarction, and stroke.
2. General surgical complications of such major surgery include bleeding (requiring transfusion) and wound infection(s).
3. Complications pertinent to transhiatal esophagectomy include ischemic necrosis of the gastric conduit, with or without mediastinitis or empyema; anastomotic leak, with or without mediastinitis or empyema; left recurrent laryngeal nerve injury; injury to the thoracic duct with ensuing chylothorax; splenic injury necessitating splenectomy; and pleural injury causing pneumothorax and/or pleural effusion.
4. Potentially catastrophic intraoperative injuries include those to the membranous airway or major vessels (e.g., aorta, azygos vein).

Follow-Up

1. Postoperative follow-up at 2-3 weeks entails clinical examination, chest X-ray, and removal of the jejunostomy catheter.
2. If postoperative adjuvant therapy (chemotherapy, radiation therapy) is recommended, it can be started at about 4 weeks postoperatively.
3. Subsequent follow-up is at 3-month intervals and includes clinical examination and chest X-ray. If esophagectomy was performed for carcinoma, follow-up must be conducted for a total of 5 years after conclusion of treatment. CT scans may be obtained to monitor for recurrence at 6-12 month intervals, assuming that clinical progress is otherwise favorable.

Chapter 78

Mediastinoscopy: Cervical

James W. Frederiksen

Indications

The objectives of cervical mediastinoscopy are to establish a histologic diagnosis of level 2, 4, and 7 lymph nodes and to assess the operability of central lung cancers. The indication for the operation is middle mediastinal adenopathy suspected from clinical history, chest X-ray, or CT scan. The more common etiologies of middle mediastinal adenopathy include primary malignancies, e.g., primary bronchogenic carcinoma; granuloma-producing diseases of infectious etiology, such as *Mycobacterium tuberculosis* and *Histoplasma capsulatum*; and granuloma-producing diseases of uncertain etiology, such as sarcoidosis.

Preop

A patient undergoing cervical mediastinoscopy requires no specific preoperative preparation. The operation is usually performed with a concomitant bronchoscopic examination, which is performed immediately prior to the cervical mediastinoscopy.

Contraindications to cervical mediastinoscopy include proven or suspected ascending or arch aortic aneurysm or dissection, and severe cervical spine disease that limits adequate neck extension.

Procedure

Step 1. The patient is placed in the supine position on the operating table. The neck is extended. An inflatable support placed beneath the shoulders facilitates neck extension.

Step 2. Sterile drapes are applied to expose the anterior aspect of the neck and the upper chest.

Step 3. A 1.5-2.5 cm transverse curvilinear skin incision is made in the suprasternal notch.

Step 4. The incision is carried deep in the midline as the strap muscles are separated with blunt dissection. Intermittent palpation of the trachea helps to guide the surgeon directly toward that structure. Occasionally division of a few millimeters of the inferior portion of the thyroid gland is required as the surgeon approaches the trachea.

Step 5. The pretracheal fascia is divided.

Step 6. The tip of an extended index finger is placed on the exposed tracheal rings and is advanced toward the carina along and in direct contact with the more distal tracheal rings. When fully inserted, the finger will occupy a position between the trachea and the ascending aorta-innominate artery junction. The arterial pulsation should be readily felt when the finger is in this position. In addition, level 2-R and level 4-R lymph nodes are occasionally palpable with the tip of the finger.

Northwestern Handbook of Surgical Procedures, edited by Richard H. Bell, Jr. and Dixon B. Kaufman. ©2005 Landes Bioscience.

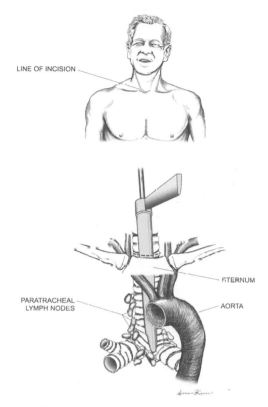

LINE OF INCISION

STERNUM

PARATRACHEAL
LYMPH NODES

AORTA

Figure 78.1. Cervical mediastinoscopy.

Step 7. After the finger is removed, the beveled end of the mediastinoscope is placed over the exposed upper tracheal rings, which are visualized through the scope.

Step 8. The mediastinoscope is advanced under direct vision along the anterior tracheal wall into the same space created by the previously inserted finger.

Step 9. Through the mediastinoscope, level 2, 4, and 7 lymph nodes are identified by blunt dissection with a small-caliber rigid suction dissector whose rim is equipped with electrocautery.

Step 10. Cup-tipped or alligator-tipped biopsy instruments are passed down the lumen of the mediastinoscope and are used to biopsy dissected lymph nodes.

Addsteps. Occasionally the appearance of the azygous vein or the superior vena cava will mimic the appearance of an anthracotic lymph node. Biopsy of either venous structure can produce massive bleeding. This complication usually can be prevented by inserting a 22-gauge spinal needle into the putative lymph node and aspirating. Biopsy is not performed if blood is aspirated. If the wall of a large vein is biopsied, the bleeding is usually controlled by firmly packing the pretracheal space with Surgicel and a gauze sponge for 10-30 minutes. The same is true for injuries to the pulmonary artery since this is also a low-pressure system. If no bleeding occurs following removal of the sponge, the mediastinoscopy incision is closed. If bleeding persists, the pretracheal space is repacked, and the bleeding is controlled via either a median sternotomy or a right thoracotomy. The incision used is determined by the patient's pathological anatomy and clinical status.

Postop

The patient usually can be extubated in the operating room or shortly after transfer to the recovery room. A chest X-ray is performed in the recovery room to assess the mediastinum and whether a pneumothorax resulted from the dissection. If no intra- or postoperative complication occurs, the patient usually can resume her preoperative diet within an hour or two and usually can be discharged several hours later.

Complications

Please see "Addsteps" above.

Follow-Up

An outpatient office visit that includes a PA and lateral chest X-ray is scheduled 1-2 weeks following the operation.

Chapter 79

Lung Biopsy: Thoracoscopic

James W. Frederiksen

Indications

The objectives of the operation are:

1. to establish a histologic diagnosis in a patient with clinical and/or X-ray evidence of interstitial lung disease or diffuse air space disease;
2. to establish a histologic diagnosis of a peripheral lung nodule; and
3. to obtain lung tissue for culture of aerobic and anaerobic organisms, acid-fast organisms, fungi, viruses, and parasites in a patient with a suspected pulmonary infection.

The indications for the operation include suspected interstitial lung disease, peripheral lung nodule(s) identified on chest X-ray or thoracic CT scan, and suspected pulmonary infection in a patient in whom culture or other assessment of sputum, tracheal aspirate, bronchoalveolar lavage fluid, or transbronchial biopsy is unlikely or has failed to yield a diagnosis.

Preop

The surgeon should carefully assess the ventilatory status and the coagulation function of each patient in whom the operation is contemplated. The operation is contraindicated in persons likely to become hypoxic during single-lung ventilation.

Significant prolongation of the PT (INR > 1.5), any prolongation of the PTT, and significant thrombocytopenia (platelet count < 50,000) increase the risk of serious intra- and postoperative bleeding. Coagulopathic patients who require thoracoscopic lung biopsy may need pre- and intraoperative transfusions of fresh frozen plasma and platelets.

Procedure

Step 1. The patient is anesthetized while lying supine. Single-lung ventilation capability is achieved either through use of a double-lumen endotracheal tube or insertion of a bronchial blocker. Subsequently the patient is turned to the lateral decubitus position. Flexing the table slightly downward prevents the ipsilateral hip from limiting full excursion of the video camera.

Step 2. The hemithorax is draped in a manner suitable for a full thoracotomy. Rarely during a thoracoscopic biopsy operation, hemorrhage that is not controllable with thoracoscopic instruments occurs. Hence, the operative field should be prepared and draped widely enough for a thoracotomy, and instruments required to perform a thoracotomy should be readily available during this and all other thoracoscopic operations.

Northwestern Handbook of Surgical Procedures, edited by Richard H. Bell, Jr. and Dixon B. Kaufman. ©2005 Landes Bioscience.

79

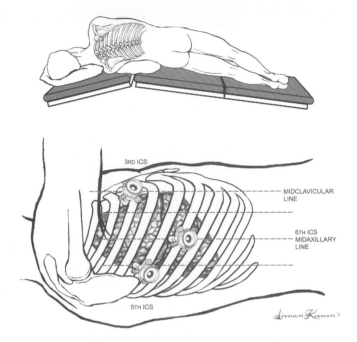

Figure 79.1. Thoracoscopic lung biopsy. Patient position and trocar placement.

Step 3. The anesthesiologist allows the lung to deflate as the surgeon creates three 2 cm thoracoscopic port incisions: one in the 6th, 7th, or 8th intercostal space in the midaxillary line, one in the 3rd or 4th intercostal space between the midclavicular line and the anterior axillary line, and one in approximately the 5th intercostal space posterior to the lateral border of the scapula.

Step 4. The surgeon inserts the video camera—thoracoscope—through the midaxillary line port.

Step 5. The surgeon inserts other endoscopic instruments, e.g., lung clamps, retractors, scissors, electrocautery, and later, a stapler, through the other two ports. Most endoscopic staplers are designed so that each firing creates four closely spaced parallel rows of staples. A blade that passes between the two middle rows of staples subsequently divides the tissue between these rows.

Step 6. The surgeon thoroughly examines the lung, the pleura, and the mediastinum with the thoracoscope. To facilitate the examination, it is often necessary to retract the lung with endoscopic clamps, to divide adhesive bands with endoscopic scissors or electrocautery, and to insert the thoracoscope through one or both of the more superior port incisions. Occasionally the surgeon can feel a small portion of the lung surface by inserting a finger through a port incision.

Step 7. Guided by preoperative X-ray studies and by visual inspection, the surgeon identifies areas of abnormal and normal appearing lung.

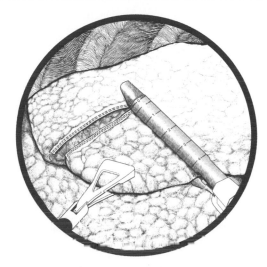

Figure 79.2. Thoracoscopic lung biopsy. Wedge resection.

Step 8. To establish a histologic diagnosis of a diffuse disease process, or to obtain tissue for culture, the surgeon usually excises a small "wedge" of abnormal appearing lung from each lobe with an endoscopic stapling instrument. The surgeon holds steady the lung tissue intended for removal by applying an endothoracic clamp or a ring forceps to that tissue as the stapler is being positioned and fired. If normal-appearing lung tissue is present, the surgeon should include some of this tissue in at least one of the specimens.

Step 9. The surgeon sends at least one specimen containing abnormal lung tissue to the pathologist for frozen section examination. A brief summary of the patient's history and the suspected diagnosis should accompany the specimen so that the pathologist can prepare the tissue for special stains or for electron microscopic examination if appropriate. Especially if the patient is immunocompromised, or if the referring pulmonologist or infectious disease specialist suspects an infectious process, the surgeon sends tissue samples for appropriate cultures and bacteriologic stains.

Step 10. If one or more pulmonary nodules are to be biopsied, the surgeon confirms the presence of the nodule by visual inspection.

Step 11. With a stapling device the surgeon excises the nodule, along with a surrounding 1- to 2-cm margin of normal lung. The surgeon then inserts the specimen into a plastic bag that is equipped with a purse string. The bag is closed by tightening the purse string before the bag is removed through a port incision.

Step 12. Following removal of the biopsy specimen(s), the surgeon makes certain that each staple line has no significant air leak or bleeding. The surgeon inserts a large-bore polyethylene cannula via the midaxillary line port incision and advances the tip to the apex of the hemithorax. Visualizing the chest tube's insertion and verifying its apical position can be facilitated by inserting the thoracoscope via one of the more superior port incisions as the polyethylene catheter is inserted.

Step 13. The anesthesiologist inflates the lung as the surgeon secures the polyethylene catheter with a heavy suture and closes the other port incisions with absorbable suture. The chest tube is connected to a PleurEvac suction device set at 20 cm water suction.

Addsteps. RE: Step 9. The pathologist is asked to examine a frozen section of a biopsy specimen, not necessarily to make a definitive diagnosis but to be certain that a diagnosis can be made from permanent sections.

RE: Step 11. A wedge resection performed to obtain lung tissue for histologic examination or for culture can occasionally be accomplished with a single firing of the stapling instrument. However, wedge resection of a pulmonary nodule usually requires at least two, and sometimes more, firings from multiple sites adjacent to the nodule. A resected specimen that contains a suspected malignant nodule is removed in a closed plastic bag via a port incision to help prevent malignant cells from being deposited along the incision during removal.

Postop

The patient usually can be extubated in the operating room or shortly following transfer to the recovery room. A chest X-ray is obtained in the recovery room to assess the expansion of the lungs and the position of the polyethylene catheter. The PleurEvac device usually is maintained at 20 cm water suction for 24 hours. Thereafter suction is discontinued, as long as the patient's lung has become and remained fully expanded. The polyethylene catheter is removed when significant drainage has ceased and when the lung has become satisfactorily expanded.

Complications

Bleeding, prolonged air leak, and respiratory failure are the most common early postoperative complications.

Follow-Up

An outpatient visit that includes a chest X-ray is scheduled for approximately 2 weeks following operation.

Chapter 80

Pulmonary Lobectomy: Open

David Fullerton

Indications

The indications for pulmonary lobectomy include:

1. removal of lung cancer;
2. removal of lung destroyed by infection, especially tuberculosis; and
3. removal of lung abscess refractory to medical therapy.

Preop

Preoperative assessment is focused on whether or not the patient has the pulmonary reserve to withstand removal of the lobe. Two tests are used for this purpose: pulmonary function tests (PFTs) and the arterial blood gas (ABG). Based upon preoperative PFTs, the patient must be predicted to have a postoperative FEV1.0 of at least 800 cc. A preoperative ABG demonstrating an arterial pCO_2 greater than 40 is a contraindication to resection. It suggests insufficient alveolar ventilation to support the patient postoperatively.

Procedure

Step 1. The procedure is performed under general anesthesia. The anesthesiologist intubates the patient with a double-lumen endotracheal tube or uses a bronchial blocker to permit one-lung ventilation during the procedure.

Step 2. The patient is placed in the lateral decubitus postion. The knees are appropriately padded to prevent injury to the peroneal nerves. A rolled towel is placed under the axilla to prevent pressure injury to the axillary structures. The ipslateral arm is supported with an airplane splint to permit exposure to the posterolateral chest wall.

Step 3. The incision is initiated posteriorly, midway between the medial scapular border and the spinous processes. It extends anteriorly, in a curvilinear fashion, two fingerbreadths below the scapular tip. The incision extends to approximately the anterior line, two fingerbreadths below the level of the nipple.

Step 4. The muscles of the chest wall are divided with electrocautery. The 5th intercostal space is entered.

Step 5. The lung is deflated as the anesthesiologist employs one-lung ventilation. The lung is reflected posteriorly.

Step 6. The pulmonary artery is identified in the pulmonary hilum. The pulmonary arterial blood supply to the appropriate lobe is sharply dissected out and the branches ligated in continuity and divided.

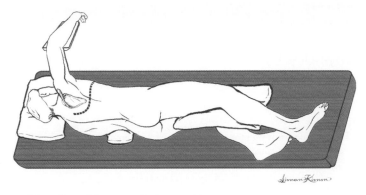

Figure 80.1. Open lobectomy. Patient position.

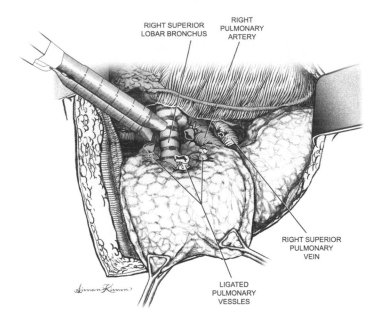

Figure 80.2. Open lobectomy. Dissection.

Step 7. The pulmonary venous drainage is identified: upper lobes drain into the superior veins and lower lobes drain into the inferior veins. The appropriate vein is sharply dissected out, ligated in continuity, and divided. The vein is oversewn with 4-0 polypropylene suture.

Step 8. The lobar bronchus is sharply dissected out and stapled with an automatic stapling device. The bronchus is divided distal to the staple line and the lobe removed.

Step 9. The anesthesiolgist is requested to inflate the remaining lung and to hold inflated under 20 cm H_2O pressure. The bronchial staple line is submerged in saline, and an air leak ruled out.

Step 10. Anterior and posterior chest tubes are placed.

Step 11. The intercostal space is closed with paracostal sutures. The musculature of the chest wall, subcutaneous tissue, and skin are then closed in appropriate layers.

Postop

Patients should be extubated in the operating room. The chest tubes should be kept on suction and removed when the patient has no air leak and the tubes are draining less than 100 cc per 24 hours.

Complications

1. Prolonged air leak
2. Bronchial stump dehiscence
3. Empyema
4. Wound infection
5. Pneumonia

Follow-Up

Patients undergoing resection for lung cancer need lifelong follow-up with an nual chest X-rays.

Chapter 81

Pneumonectomy

Robert Vanecko

Indications

Carcinoma of the lung involving either the main stem bronchus, the main pulmomary artery, or multiple lobes of the lung comprise the indications for pneumonectomy. Occasionally pneumonectomy is indicated for chronic longstanding infectious process that essentially destroyed the functional capacity of one lung.

Preop

General endotracheal anesthesia with double lumen tube or bronchial blocker, epidural catheter, central venous line, arterial line, Foley catheter.

Procedure

Step 1. Appropriate lateral decubitus position.

Step 2. Posterior lateral thoracotomy entering the chest through the 5th or 6th intercostal space.

Step 3. Free any adhesions between the lung and chest wall and assess resectability.

Step 4. Deflate lung and retract it posteriorly and inferiorly to expose the hilum.

Step 5. Incise the mediastinal pleura anteriorly around the hilum.

Step 6. Identify the pulmonary artery and superior pulmonary vein.

Step 7. Dissect down to the adventitia of the pulmonary artery and mobilize and encircle it.

Step 8. Ligate the main trunk of the pulmonary artery and its first branches, then transect the artery. Alternatively the artery may be divided between vascular staples.

Step 9. Identify the pulmonary vein, mobilize and either ligate or staple it and transect the vein.

Step 10. Displace the lung anteriorly and superiorly to expose the pulmonary ligament.

Step 11. Incise the pulmonary ligament from the diaphragm towards the hilum, exposing the inferior pulmonary vein.

Step 12. The inferior pulmonary vein is moblized, ligated or stapled and divided.

Step 13. The main bronchus is identified and lymph nodes are mobilized towards the lung.

Step 14. The bronchus is stapled and transected.

Step 15. Irrigate the pleural cavity with warm saline. Confirm airtight closure of bronchial stump.

Addsteps. If necessary reinforce bronchial closure with pleural flap. Close the chest without a chest tube.

Northwestern Handbook of Surgical Procedures, edited by Richard H. Bell, Jr. and Dixon B. Kaufman. ©2005 Landes Bioscience.

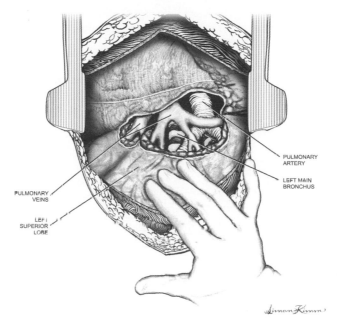

Figure 81.1. Pneumonectomy. Anatomy.

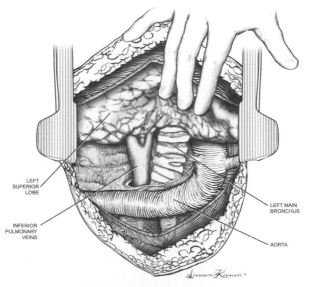

Figure 81.2. Pneumonectomy. Anatomy.

Postop

Monitor oxygen saturation, position of the mediastinaum, and level of accumulation of fluid within the pleural space.

Complications

Complications include disruption of bronchial closure, infection within pleural space, mediastinal shift with cardiovascular instability, arrhythmias.

Follow-Up

Serial chest X-rays until pleural space is stable.

81

Chapter 82

Pleurodesis: Thoracoscopic

David Fullerton

Indications

The purpose of the operation is to obliterate the pleural space to prevent reaccumulation of fluid or air. The indications are for treatment of recurrent spontaneous pneumothorax and malignant pleural effusion.

Preop

The procedure is performed under general anesthesia. Single-lung ventilation is accomplished with either a double-lumen endotracheal tube or a bronchial blocker. The patient is placed in the lateral decubitus position. The knees are padded to prevent peroneal nerve injury. A rolled towel is placed under the axilla to prevent pressure injury to the axillary structures. The ipsilateral arm is suspended on an airplane splint.

Procedure

Step 1. Two 2 cm thoracoscopic port incisions are made, one in the 5th intercostal space near the lateral border of the scapula and one in the midaxillary line in the 6th intercostal space.

Step 2. The thoracoscope is inserted into the midaxillary port. A long clamp holding a folded sponge is inserted through the other port.

Step 3. The parietal pleural surface along the chest wall is mechanically roughed-up with the folded sponge. Particular emphasis is placed on roughing up the apex of the thoracic space. Care is taken to avoid injury to the subclavian vessels and brachial plexus.

Step 4. After the mechanical pleurodesis has been completed, a solution of 100 ml of doxycycline is placed into the pleural space to add a chemical pleurodesis.

Step 5. A chest tube is placed through the midaxillary port incision.

Step 6. The posterior port is closed in appropriate layers.

Postop

The patient should be extubated in the operating room. The chest tube remains in place until the patient has no air leak, and the chest tube drainage has stopped.

Complications

1. Pneumonia
2. Wound infection
3. Empyema

Northwestern Handbook of Surgical Procedures, edited by Richard H. Bell, Jr. and Dixon B. Kaufman. ©2005 Landes Bioscience.

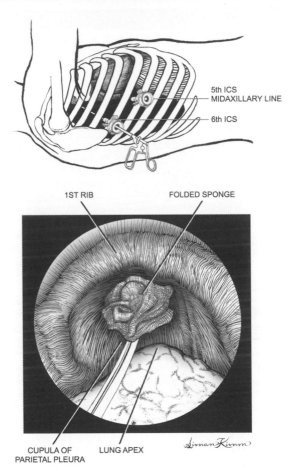

82

Figure 82.1. Thoracoscopic pleurodesis.

Follow-Up

Chest X-rays should be followed for approximately the first 2 months after the procedure. Thereafter, chest radiography is indicated for recurrent symptoms.

Chapter 83

Tracheostomy

Keith A. Horvath

Indications

The tracheostomy operation has been performed since antiquity for the emergency management of upper airway obstruction. Presently the principal indications are:

1. relief of upper airway obstruction;
2. control of secretions in patients prone to aspirate; and
3. long-term ventilatory support for patients with respiratory failure.

Tracheostomies most commonly are performed for respiratory insufficiency requiring prolonged mechanical ventilation. Typically, these patients have had an endotracheal or nasotracheal tube prior to their tracheostomy. While there is no firm rule regarding the length of time a patient may be managed with an endotracheal tube, tracheostomies are usually performed after 10-14 days of intubation in patients who are unlikely to be extubated. Originally, this time frame was shorter due to the potential for tracheal stenosis from the circumferential cuff on an endotracheal tube. While this complication has decreased significantly with the use of low-pressure cuffs, there is evidence that long-term endotracheal intubation can lead to laryngostenoses. This can be avoided by tracheostomy.

Preop

While tracheostomy may be done with local anesthesia, with the patient in a supine position and the neck hyperextended, due to the fact that an endotracheal tube is in place, it is typically done with a general inhalational anesthetic. While the procedure may be done at the bedside, it should be done in the operating room for sterility and to allow the operator the best visibility and reduce complications.

Procedure

Step 1. With the patient supine and the neck hyperextended, the patient is sterilely prepped and draped.

Step 2. Palpation of the neck reveals the position of the cricothyroid membrane with the cricoid cartilage below it.

Step 3. A horizontal incision, approximately 3-5 cm, is placed at the level of the second tracheal cartilage and is carried through the platysmal muscle.

Step 4. The strap muscles may be separated vertically in the midline to avoid bleeding.

Step 5. The thyroid isthmus is identified and dissection is carried out below it and below the lower border of the cricoid cartilage.

Step 6. The thyroid isthmus may need to be divided and if so, hemostasis is critical. Mattress sutures on either side can assist with retraction and hemostasis.

Northwestern Handbook of Surgical Procedures, edited by Richard H. Bell, Jr. and Dixon B. Kaufman. ©2005 Landes Bioscience.

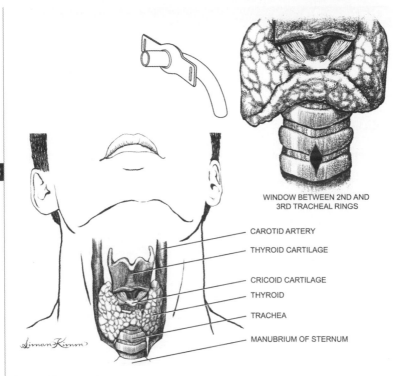

WINDOW BETWEEN 2ND AND
3RD TRACHEAL RINGS

CAROTID ARTERY

THYROID CARTILAGE

CRICOID CARTILAGE

THYROID

TRACHEA

MANUBRIUM OF STERNUM

Figure 83.1. Tracheostomy.

Step 7. More commonly, the thyroid can be dissected free and retracted cranially.

Step 8. The cartilaginous rings must be identified and the exact number of the ring must be double-checked.

Step 9. It is imperative that the first cartilage be left intact so that pressure by the tracheostomy tube will not erode the first ring or the cricoid cartilage.

Step 10. Traction sutures should then be placed through the membranous portion around the 3rd cartilage laterally. Typically, this is a 3-0 or 4-0 monofilament suture.

Step 11. The 2nd and 3rd cartilages, and potentially part of the 4th cartilage, if necessary, are incised at the midline.

Step 12. Small flaps may be created to enlarge the opening, but care should be taken to avoid making too large an opening, even with a flap that is left as a "trap door." Any opening will heal by scarring, and the larger the opening the more likely it is for the ensuing scar to narrow the trachea.

Step 13. Care should be taken to avoid damage to the cuff of the endotracheal tube when incising the trachea. The endotracheal tube should be visible through the opening and the cuff should clearly be below the stoma site.

Step 14. Once the tracheostomy opening has been created, the cuff on the endotracheal tube is deflated and the tube is removed.

Step 15. Under direct vision, using the traction sutures, the tracheostomy tube is then slid into position.

Step 16. The cuff of the tracheostomy tube should then be inflated and checked to make sure that it is intact.

Step 17. The tracheostomy may then be connected to the anesthesia/ventilator circuit.

Step 18. Exhaled carbon dioxide should be noted once connected to insure correct placement.

Step 19. The platysma and skin can then be closed with the traction sutures carried out to the skin level to provide easy access to the stoma site, should the tracheostomy tube become dislodged.

Step 20. A gauze dressing is placed at the level of the skin to protect the skin from the external flange and collar that helps hold the tracheostomy tube in place.

Step 21. This flange should initially be sutured to the skin, but once a tract is formed, the sutures may be removed.

Postop

An arterial blood gas and a chest X-ray should be performed to insure the proper functioning and position of the tracheostomy tube. Additional care is similar to an endotracheal tube with regard to suctioning to control secretions and maintain airway patency.

Complications

There are principally three long-term complications of tracheostomy: infection; hemorrhage; and airway obstruction. Additional complications include tracheoesophageal fistula or persistence of the stoma. As would be expected, the longer a tracheostomy is in place, the more likely it is that complications will occur.

1. *Infection.* All tracheostomies are contaminated and frequently will grow numerous bacteria. Therefore, sterile care and cleansing of the stoma site and appropriate maintenance of the respiratory equipment is necessary to minimize the possibility of a lower airway infection.

2. *Hemorrhage.* Occasionally, the tracheostomy tube may erode into the innominate artery and massive hemorrhage can occur. There may be a brief, significant bleed that signals the fistulization. In such a case, immediate tamponade of the arterial leak by finger pressure and prompt surgical treatment are required. The tracheostomy may need to be removed and an endotracheal tube placed. However, as this is done in an emergency setting, care must be taken to avoid loss of control of the airway. The injured artery should be resected and both ends sutured closed. Prosthetic graft material should not be used in this contaminated field. The tracheal innominate fistula is fortunately a rare, but frequently a lethal complication. In the few patients who have been successfully treated, neurologic sequela have not been noted as a result of resecting and suturing the innominate artery. Bleeding from granulation tissue or the skin is far more common but is usually less massive and can be treated with local measures.

3. *Airway obstruction.* Despite having a tracheostomy tube in place, airway obstruction can occur for a variety of reasons. The cuff may prolapse over the end of the tracheostomy due to overdistention, but this should be avoidable. Crusting of secretions and obstruction of the tube can be addressed by the use of an inner cannula, humidification, and suctioning. Occasionally, granulations may form at the end of the tracheostomy, which may need to be removed.

The most troubling type of obstruction is due to tracheal stenosis. This can be prevented by having lightweight connectors that allow for pivoting around the connection site, as opposed to torquing on the tracheostomy. Creation of an appropriate-sized stoma will prevent stenosis later as the stoma heals. Large-volume, low-pressure cuffs that are typically employed, conformed to the shape of the trachea rather than distending it, are key to preventing pressure necrosis and ensuing tracheal stenosis.

Follow-Up

Long-term follow-up is primarily removal of the tracheostomy once the patient's respiratory failure has abated. As a temporary measure, a cap or plug can be placed to keep the stoma open while the patient breathes through their mouth. Prior to this, a fenestrated tracheostomy will allow air to pass across the vocal cords and permit the patient to talk. Once mechanical ventilation is no longer needed, the cap or plug may be removed and the stoma site will heal. Typically, no additional procedures are needed to promote this process.

83

Section 6: Transplantation

Section Editor: Dixon B. Kaufman

Chapter 84

Arteriovenous Graft (AVG)

Alan J. Koffron

Indications

1. Long-term hemodialysis access;
2. inability to perform continuous ambulatory peritoneal dialysis (CAPD); and
3. inability to place arteriovenous fistula.

Preop

Routine preoperative screening with special emphasis on cardiovascular system. Use of the nondominant upper extremity is preferred. Preoperative vascular assessment should include arterial and venous evaluation: arterial (Allen's test, sonography), venous (sonography, venography).

Procedure

Step 1. Supine position with target extremity abducted on armboard.

Step 2. Circumferential sterile prep/surgical drape from finger tips to lateral border of pectoralis major.

Step 3. Antecubital fossa: Mark skin over the palpated brachial artery, and, if visible, antecubital veins.

Step 4. Incision: 2 cm distal to the antecubital fold, hemostasis using electrocautery and absorbable suture material.

Step 5. Venous dissection: Target vein (cephalic or median antecubital vein) isolated and division of vein branches avoided (maintain venous drainage of extremity and preserve veins for later use in secondary AVG or revisions of primary AVG). Obtain vascular control for about 3-4 cm.

Step 6. Arterial dissection: Divide the subcutaneous tissue down to the biceps aponeurosis. Divide the aponeurosis sharply in a cruciate formation for optimal exposure to the artery and concomitant veins below. Palpate the brachial artery, separate concomitant veins (on each side of artery), and avoid ligation of veins if possible. Arterial dissection is greatly simplified by dissecting in the plane of Leriche (periadventitial plane) for about 3 cm for vascular control. Vessel loops are placed around vascular structures to assist in control prior to anastomosis.

Step 7. Graft selection: Thin-walled 4-7 mm tapered polytetrafluoroethylene (PTFE) grafts are preferred as the small (4 mm) end is anastomosed to the arterial inflow to avoid postoperative "steal" syndrome. Use graft length ensuring redundancy so that kinking and tension are avoided.

Step 8. Subcutaneous tunnel: Select a semicircular tunnel device (Noon, Kelly-Weck, or sheath) of a diameter slightly greater than that of the selected PTFE graft. Advance the tunnel device from the antecubital incision distally along the medial/volar aspect of the forearm. Create a small cutdown incision distally and then pass

Northwestern Handbook of Surgical Procedures, edited by Richard H. Bell, Jr. and Dixon B. Kaufman. ©2005 Landes Bioscience.

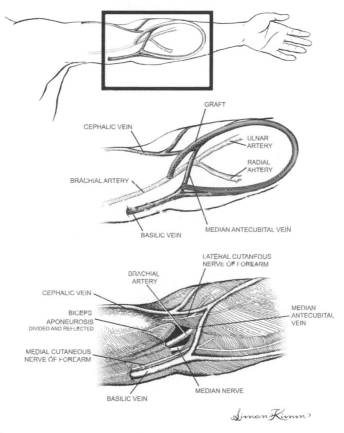

GRAFT

CEPHALIC VEIN

ULNAR ARTERY

RADIAL ARTERY

BRACHIAL ARTERY

BASILIC VEIN

MEDIAN ANTECUBITAL VEIN

LATERAL CUTANEOUS NERVE OF FOREARM

BRACHIAL ARTERY

CEPHALIC VEIN

MEDIAN ANTECUBITAL VEIN

BICEPS APONEUROSIS DIVIDED AND REFLECTED

MEDIAL CUTANEOUS NERVE OF FOREARM

BASILIC VEIN

MEDIAN NERVE

Figure 84.1. Arteriovenous graft (AVG).

the PTFE graft (retrograde) through the tunnel. Repeat these steps on the lateral/volar aspects of the forearm to create an oval graft tunnel where the tapered (4 mm) end of the graft lies next to the brachial artery and the redundant end of the graft lies next to the target vein. Attention should be given to avoid twisting or kinking of the graft as it is passed through the tunnel. Irrigation of the graft with saline will ensure its patency prior to anastomosis.

Step 9. Venous anastomosis: Occlude the vein (and any tributaries) with atraumatic vascular clamps (e.g., Bulldog or Heifets clamps) so that sutures cannot be caught in the clamps. Incise (11 blade) the vein at the intended site of anastomosis; irrigate with heparinized saline. Extend the venotomy with Dietrich's scissors to a length to match the graft, estimating the angle at which the graft will be cut. Tailor the graft to match venotomy and anastomose (end-to-side, or end-to-end) using fine (5-0 or 6-0) nonabsorbable monfilament suture (polypropylene, etc.). Vascular control is relieved and blood is allowed to flow back through the venous anastomosis; the graft is flushed with heparinized saline and occluded with a soft vascular clamp.

Step 10. Arterial anastomosis: Obtain vascular control (atraumatic clamps) proximal and distal to the site of anastomosis. Create a small arteriotomy (11 blade) with care to avoid injuring the posterior arterial wall. Flush with heparinized saline. Extend the arteriotomy to approximately 6 mm (4 mm graft end will be cut at a slight angle, matching the arteriotomy site). Perform anastomosis (as done with the vein) with attention to avoid intimal injury. Incorporate all layers of the arterial wall to prevent intimal dissection/anastomotic bleeding.

Step 11. Graft perfusion: Remove the graft clamp (outflow) and remove the distal arterial clamp (back bleed). Then remove the proximal arterial clamp. There should be brisk flow, often felt as a thrill through the graft and the vascular anastamosis sites.

Step 12. Hemostasis: Avoid unnecessary manipulation of the anastomosis. Surgical gauze with gentle pressure will stop needle-hole oozing.

Step 13. Closure: The wound is closed with 3-0 or 4-0 absorbable subcutaneous sutures; skin is closed with 4-0 or 5-0 subcuticular sutures.

Step 14. Dressing: Loose gauze. Avoid circular or semicircular dressings or tapes (patients risk postoperative swelling and dangerous tourniquet effect).

Postop

Use arm freely; elevate arm while at rest. Oral analgesics.

Complications

Early complications: AVG thrombosis (urgent thrombectomy/revision), postoperative swelling (surveillance for infection), bleeding (reoperation, hemostasis), skin infection (antibiotics), graft infection (removal), arterial steal (graft revision).

Follow-Up

Avoid puncturing AVG for 8 weeks to allow time for tissue incorporation of PTFE in order to prevent needle-hole perigraft hematoma and graft thrombosis and reduce risk of graft infection due to repeat access.

Chapter 85

Primary Radial Artery-Cephalic Vein Fistula for Hemodialysis Access

Joseph R. Leventhal

Indications

The indication for primary radial artery-cephalic vein fistula is end stage renal disease requiring long-term hemodialysis access. Ideally, fistula creation should precede the need for hemodialysis by several months in order to allow for adequate maturation before use. Careful physical examination should be performed to rule out arterial insufficiency in the upper extremity intended for use. Examination of the cephalic vein under tourniquet should be performed to ensure the absence of stenosis or thrombosis in the forearm. Patients with a history of previous central lines, dialysis catheters, and neck/chest trauma should be evaluated to rule out central venous obstruction.

Preop

Administer prophylactic antibiotics. Monitored anesthesia care with supplemental local analgesia should be used. Do not use local with epinephrine since it causes vasospasm. Position the patient supine, with the outstretched upper extremity placed on an armboard. Prep the arm from finger tips to chest wall. Place a stockinette over the hand and drape the patient to keep the axilla exposed—in case more proximal dissection of a fistula or graft placement is required.

Procedure

Step 1. Mark the course of the radial artery and cephalic vein in the distal half of the forearm, and mark the proposed longitudinal skin incision midway between the two. If possible, identify the common dorsal branch of the cephalic vein. The use of this branch provides a "patch" anastomosis which is optimal from a hemodynamic standpoint and also technically easier to perform.

Step 2. Infiltrate the proposed incision with 1% lidocaine solution. Using a 15 blade create a 5-6 cm incision midway between the artery and vein. The distal limit of the incision should be proximal to the radial styloid process.

Step 3. Dissect through subcutaneous tissue using blunt (mosquito hemostat) techniques to expose the cephalic vein. Carefully divide tissues using scissors or electrocautery. Avoid grasping the vein proper; isolate the vein and place a vessel loop around it to aid in retraction exposure. Isolate the vein for a length of 5-6 cm. Carefully ligate and divide small branches of the cephalic vein.

Northwestern Handbook of Surgical Procedures, edited by Richard H. Bell, Jr. and Dixon B. Kaufman. ©2005 Landes Bioscience.

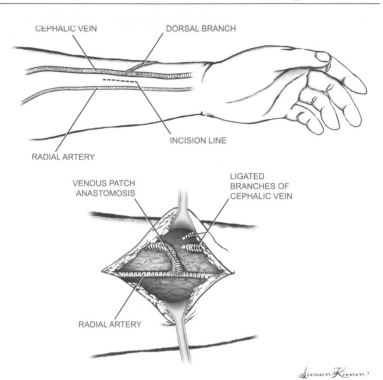

Figure 85.1. Radial artery-cephalic vein fistula.

Step 4. Dissect out and isolate the radial artery using techniques described for venous dissection. Take care to separate the concomitant veins which run on either side of the artery. Ligate or cauterize small arterial branches of the main radial artery to allow for complete mobilization. Use vessel loops to retract and move the artery. Avoid grasping the radial artery proper.

Step 5. Ligate the cephalic vein distally. Open the cephalic vein just proximal to the ligature and flush the vein with heparinized saline (5 ml of 1000 units/cc in 500 ml of saline). Apply a small bulldog clamp to the vein to prevent backflow. Using the dorsal branch of the vein, create a vein patch for anastomosis to the artery. Alternatively, spatulate open the main cephalic vein to allow for a generous anastomosis.

Step 6. Using vessel loops or bulldog clamps, occlude the radial artery proximal and distal to the proposed arteriotomy site.

Step 7. Make a small (1-2 mm) arteriotomy in the radial artery. Irrigate the artery with heparinized saline. Extend the arteriotomy to a length of 8-10 mm using Potts or Dietrich scissors.

Step 8. Sew the cephalic vein to the radial artery in an end-to-side fashion using 7-0 polypropylene suture.

Step 9. Remove the bulldog clamp from the vein. Release the distal artery, then the proximal artery vessel loop/vascular clamp.

Step 10. Establish hemostasis using gentle pressure at the anastomosis. Inspect the anastomosis and course of the cephalic vein. Be sure the vein is not kinked or twisted as it courses proximally. Feel for a thrill in the vein proximal to the anastomisis.

Step 11. Close the incision with absorbable suture.

Postop

Elevate the arm to prevent hand/arm swelling. Remove the dressing 24 hours postoperatively. Begin exercises to mature fistula (squeezing a ball several times a day for 10 minutes).

Complications

Complications include bleeding, thrombosis, infection, hand ischemia ("steal"), paresthesias from peripheral nerve injury during surgery.

Follow-Up

The patient should be examined at regular intervals to ensure fistula maturation before use in dialysis unit.

85

Chapter 86

Laparoscopic Donor Nephrectomy

Joseph R. Leventhal

Indications

The indication for laparoscopic donor nephrectomy is living donor renal transplantation. Donor evaluation should ensure that the renal donor has adequate functional reserve in the remaining kidney to allow donation to proceed. CT angiography or MR imaging should be performed to define renal anatomy before proceeding with donor nephrectomy.

Preop

Bowel prep with magnesium citrate is performed the night before surgery. Endotracheal intubation—avoid use of nitrous oxide to prevent bowel distension. Orogastric suction, bladder catheter drainage, prophylactic antibiotics, and antithrombotic sequential leg compression devices are routinely used. The patient is placed in the right decubitus position for left donor nephrectomy, and vice versa. Flex the operating table at a point midway between patient's iliac crest and rib cage and elevate the kidney rest in order to maximize exposure during the procedure. Prep and drape the patient to allow for—if necessary—open conversion to extended subcostal or standard flank approach if required. The operating surgeon stands facing the patient's abdomen, with the camera operator caudad. An assistant and scrub nurse are positioned opposite the surgeon. Standard laparoscopic instrumentation, along with a 30° laparoscope and ultrasonic scalpel are used. More than 95% of donor nephrectomies remove the left kidney, in order to obtain longer renal vein length. The operative procedure is described for left donor nephrectomy.

Procedure

Step 1. Insert a Veress needle in the subcostal location. Create a pneumoperitoneum of no more than 15 mm Hg.

Step 2. Introduce the laparoscope into abdomen using a 10 mm Visiport™.

Step 3. Place two 12 mm operating ports underneath the rib cage; a 5 mm port may be placed subcostally in the posterior axillary line to assist in retraction of the kidney. Port placement will vary slightly from patient to patient depending upon patient girth and length of the torso.

Step 4. Using an ultrasonic scalpel, mobilize the left colon and spleen medially, away from the kidney posteriorly.

Step 5. Open Gerota's fascia and dissect out the renal vein. Clip and divide gonadal, adrenal, and lumbar vein branches to obtain maximal length.

Step 6. Identify renal artery/arteries. Avoid overdissection of vessels to prevent vasospasm.

Northwestern Handbook of Surgical Procedures, edited by Richard H. Bell, Jr. and Dixon B. Kaufman. ©2005 Landes Bioscience.

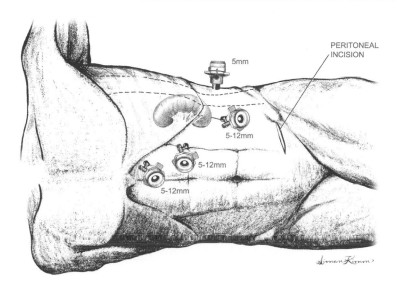

Figure 86.1. Laparoscopic donor nephrectomy. Trocar placement.

Step 7. Dissect adrenal gland off the upper pole of the kidney.

Step 8. Identify the ureter inferior to the lower pole of the kidney. Dissect adequate length of ureter allowing for easy transplantation of kidney in recipient, usually to the point where the ureter courses over iliac vessels. Avoid stripping the ureteral blood supply.

Step 9. Completely mobilize the kidney. Ensure that the kidney is completely free except for blood vessels and ureter. Avoid torsion of kidney.

Step 10. Create a 6-7 cm extraction incision in the left lower quadrant, without violation of the peritoneum.

Step 11. Anticoagulate the patient with 5000 units of heparin sodium. Distally clip and divide the ureter.

Step 12. Divide the renal artery, followed by the renal vein, with a linear vascular laparoscopic stapler. Reverse the heparin with protamine sulfate. The kidney is now completely free.

Step 13. Open the peritoneum at the extraction site and deliver the kidney through this wound by hand into an iced saline solution.

Step 14. Remove the staple lines from the donor kidney blood vessels and flush out the kidney with a chilled preservation solution (Collins, Viaspan, etc.).

Step 15. Inspect the abdomen through the extraction incision. Confirm secure placement of clips on the distal ureter. Palpate the iliac artery to confirm good pulsations.

Step 16. Close the extraction incision with #1 or 0 absorbable suture. Reestablish pneumoperitoneum and inspect the operative field laparoscopically. Confirm adequate hemostasis, remove ports under direct visualization, and desufflate the abdomen. Close all incisions.

Postop

Foley catheter drainage until postoperative day 1. Intravenous fluids at 50 cc/h until midnight, then heparin-lock IV. Parenteral analgesia until postoperative day 1, then switch to oral. Clear liquid diet the night after surgery, ambulate in the evening. Hemogram and renal chemistry panel on postoperative day 1. Anticipate discharge on postoperative day 1 or 2.

Complications

Complications include subcutaneous emphysema, atelectasis, infection, bleeding from vascular staple lines, splenic injury, and vascular injury.

Follow-Up

Follow-up in clinic first week postoperatively. Check renal function. No other office follow-up needed.

Chapter 87

Kidney Transplantation

Dixon B. Kaufman

Indications

Indications for kidney transplantation are irreversible chronic or end-stage renal disease. Contraindications include active or chronic infection, active or recent (<2 years) malignancy, active glomerulonephritis, life expectancy <1 year, sensitization to donor tissue, serious (untreatable) preexisting comorbidities, medical noncompliance, active substance abuse, uncontrolled psychiatric disorders.

Preop

General orotracheal anesthesia, Foley bladder catheter, central venous catheter (optional, usually for recipients of cadaveric grafts). Upper extremity positioning to prevent occlusion of arteriovenous graft/fistula.

Procedure

Step 1. Prone position. Right (or left) lower quadrant "hockey-stick" incision positioned two fingerbreadths medial to anterior superior iliac spine and one fingerbreadth superior to pubis.

Step 2. Incision is carried through the skin, subcutaneous tissue, and external oblique aponeurosis. Tendon-like insertion of medial edge of external/internal oblique musculature and lateral edge of rectus sheath are opened down to (but not through) the peritoneum. Inferior epigastric vessels are suture-ligated and divided. The round ligament (in females) is suture-ligated and divided. The spermatic cord (in males) is encircled with a Penrose drain and retracted medially. The peritoneum is mobilized medially and cephalad to expose the retroperitoneal iliac fossa. Mechanical retractors are often used to aid in exposure.

Step 3. Expose and mobilize external iliac vein and artery by suture ligation and division of overlying lymphatics. It is sometimes necessary to expose and mobilize the hypogastric artery and common iliac artery. To gain maximal exposure of iliac vein, suture ligate (stick-ties) and divide hypogastric vein(s). This allows the external/common iliac vein to be elevated and mobilized lateral to the iliac artery.

Step 4. The location of the vascular anastomoses depends on the geometric relationships of the length of the renal vessels (short with living donor organs), size of the renal allograft, and anterior/inferior ascent of the psoas muscle. Position the renal allograft proximal enough to lie flat in the iliac fossa without the lower pole of the kidney being excessively tipped up by the psoas muscle. Often intravenous diuretics are slowly administered (over 30 minutes) during the vascular anastomoses.

Northwestern Handbook of Surgical Procedures, edited by Richard H. Bell, Jr. and Dixon B. Kaufman. ©2005 Landes Bioscience.

87

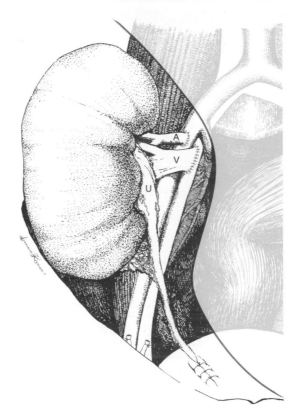

Figure 87.1. Kidney transplantation. Completion of anastomoses.

Step 5. If the patient is preuremic, adminster parenteral heparin. Uremic patients do not require heparin unless a hypercoagulable state is known. Place a vascular clamp on the iliac vein. Perform a venotomy and irrigate with heparinized saline. Anastomose the renal vein end-to-side to the iliac vein using 5-0 monofilament nonabsorbable suture.

Step 6. Place vascular clamps on the iliac artery proximally relative to the venous anastomosis. The site of the arteriotomy may be the external or common iliac artery. Alternatively, the hypogastric artery may be used by suture ligation of the distal aspect with proximal placement of a bulldog clamp (maintains iliac arterial bloodflow; end-to-end anastomosis performed). Perform an arteriotomy and irrigate with heparinized saline. Perform an end-to-side anastomosis with 6-0 monofilament nonabsorbable suture.

Step 7. Slowly release the venous clamp and then release arterial clamps to allow full perfusion of the renal allograft. Control bleeding points with suture ligation or electocautery.

Step 8. Begin ureterovesical construction by infusing sterile saline into the bladder via side-port in cysto-tubing. Clamp Foley tubing to keep the bladder distended. Properly align the ureter, measure to the appropriate length, and suture ligate and divide the distal ureteral artery. Cut and spatulate the distal ureter. In males, slip the ureter under the spermatic cord.

Step 9. (optional) Place a 6 F double-J silastic ureteral stent.

Step 10. For anterior ureteroneocystotomy, incise the bladder musculature down to (not through) the bladder mucosa. Create the cystotomy (approximately 2 cm) and perform mucosa-to-mucosa anastomosis with 5-0 monoflament absorbable suture. Create an antireflux tunnel over the distal aspect of the ureter by approximating the muscular layer with 4-0 monoflament absorbable suture as interrupted stiches.

Step 11. Irrigate the surgical field with antibiotic solution and position renal allograft to lie flat, avoiding tension on the vascular anastomosis and ureter.

Step 12. Close the incision by approximating the internal/external oblique musculature and external oblique aponeurosis in a single layer with #1 monofilament nonabsorbable suture. Approximate the skin edges with a stapler. Dress the incision.

Postop

Continue immunosuppressive medical therapy according to institutional protocols. Manage fluid replacement according to the rate of urine output and intravascular fluid status (clinical assessment ± central venous pressure measurement). Administer diuretics according to urine output and intravascular fluid balance. Routine assessment of urine output, CBC, and chemistries.

Complications

Complications include hemorrhage, vascular thrombosis, ureteral leak. Metabolic complications include delayed graft function (low urine output secondary to acute tubular necrosis), electrolyte disorders (hyperkalemia, hypocalcemia, hypomagnesemia).

Follow-Up

Immuosuppression to prevent rejection with periodic trough concentration monitoring to guide dosing. Routine measurement of serum blood urea nitrogen and creatinine to assess renal allograft function.

Chapter 88

Distal Splenorenal (Warren) Shunt

Michael Abecassis

Indications

The indication for distal splenorenal (Warren) shunt is portal hypertensive bleeding refractory to medical and endoscopic therapy.

Preop

Hemodynamic stabilization; mesenteric venous examination (Doppler ultrasound, magnetic resonance venography).

Procedure

Step 1. The patient is positioned in the supine position under general anesthesia and endotracheal intubation. A Foley catheter is inserted for monitoring urine output, and a central venous catheter is utilized for monitoring central venous pressure.

Step 2. Antiseptic prep is used from the nipple line to the groin. The body is draped in sterile fashion. Appropriate antibiotics are given prior to skin incision.

Step 3. This operation consists of two essential components. First, the portosystemic shunt itself between the splenic vein and the left renal vein and, second, the disconnection between the portal vascular bed and the shunted portion of the portal bed, i.e., the splenic venous drainage.

Step 4. A transverse upper abdominal incision is used which is parallel to the long axis of the pancreas and which extends from the right to the left costal margins. Alternatively, in patients with narrow costal margins and a long epigastrium, a long upper midline incision can also be used.

Step 5. Once the peritoneal cavity is entered and appropriate retractors are used, the lesser sac is entered through the gastrocolic omentum. This dissection is carried from the duodenum laterally to, but not including, the short gastric vessels. In the process of entering the lesser sac, the gastroepiploic vessels can be multiply ligated in order to disconnect the portal system from the venous system draining the spleen. At the point of the short gastric vessels, the interruption of these collaterals can be carried out more laterally inferior to the spleen by dividing the lienal-colic ligaments as well as possible. There are often very large collaterals in this area which can act as a sump for the splenic bed. Mobilization of the splenic flexure of the colon may be helpful and necessary in order to divide and disconnect these collaterals.

Step 6. Once the lesser sac is exposed, the inferior border of the pancreas is identified and the peritoneum overlying this inferior border is incised from the area of the superior mesenteric vein (SMV) laterally towards the tail of the pancreas. An avascular plane should be developed which allows extensive superior mobilization of the body and tail of the pancreas. When necessary, lymphatic vessels should be carefully ligated in order to prevent the potential complication of chylous/lymphatic

Northwestern Handbook of Surgical Procedures, edited by Richard H. Bell, Jr. and Dixon B. Kaufman. ©2005 Landes Bioscience.

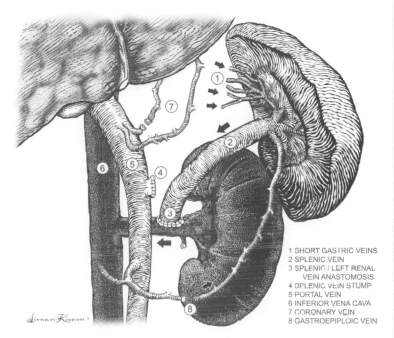

1 SHORT GASTRIC VEINS
2 SPLENIC VEIN
3 SPLENIC / LEFT RENAL
  VEIN ANASTOMOSIS
4 SPLENIC VEIN STUMP
5 PORTAL VEIN
6 INFERIOR VENA CAVA
7 CORONARY VEIN
8 GASTROEPIPLOIC VEIN

88

Figure 88.1. Distal splenorenal (Warren) shunt.

ascites postoperatively. The purpose of the pancreatic mobilization is to identify and mobilize the splenic vein. The position of the splenic vein with respect to the body and tail of the pancreas can be quite variable, and if the splenic vein is not immediately obvious, the inferior mesenteric vein (IMV) can be used as a road map to find the SMV. In some cases, the SMV may be quite superiorly placed behind the pancreas making this dissection more difficult.

Step 7. Once the splenic vein is identified, it is circumferentially dissected and mobilized. This often requires ligation and division of numerous pancreatic branches which drain the pancreas into the splenic vein. The limits of the dissection medially include the confluence of the SMV and the portal vein and laterally, division of the splenic vein into splenic branches. The IMV often enters the SMV directly or at the confluence of the SMV with the splenic vein. Regardless, it must be divided in order to afford better mobilization of the splenic vein.

Step 8. The splenic vein is mobilized along its entirety. In doing so, one may encounter the coronary vein in the medial aspect of the dissection. If it can safely be done, the coronary vein should be ligated and divided at its origin from the splenic vein. Alternatively, the left gastric vein must be divided in the lesser gastric omentum.

Step 9. Once the splenic vein is completely mobilized, our attention is turned to dissection of the left renal vein. This is done by incising the retroperitoneum posterior to the inferior aspect of the pancreas. Again, this tissue can be thick and rich in lymphatics, and great care must be taken to ligate all lymphatic tissue. The left renal vein is identified. The left adrenal and left gonadal veins are divided in order to afford better mobilization of the left renal vein. Occasionally, the IVC can be mobilized and, depending on the distance between the splenic vein and the left renal vein or IVC, either an end-to-side splenic to left renal vein anastomosis can be constructed, or the end of the splenic vein can be anastomosed directly to the side of the IVC.

Step 10. Prior to constructing the anastomosis, a vascular clamp is applied distally on the splenic vein and appropriate proximal and distal control of either the renal vein or the IVC is then achieved. The left renal vein should be used close to the IVC since it may have a functional narrowing as it crosses over the aorta.

Step 11. With appropriate vascular control, the splenic vein is cut flush almost with the SMV and the confluence with the portal vein, and this is oversewn with fine polypropylene.

Step 12. The splenic vein is then swung inferolaterally making sure that there are no kinks and no redundancy, and an end-to-side anastomosis is carried out with running fine polypropylene.

Step 13. Once the anastomosis is completed, the clamps are removed in sequence. There should be a thrill palpable over the splenic vein.

Step 14. The retractors are released and the pancreas is allowed to rest gently on the shunt. For optimal results, the splenic vein should have a large diameter and have excellent flow through it. In the case where the splenic vein is small, spontaneous splenorenal shunts may be present and, in this situation, the splenorenal shunt may not remain patent. Because portal pressure is essentially unchanged, there is no need to measure either portal pressure or portosystemic gradients following these selective shunts.

Step 15. Once the shunt is completed, a thorough examination for collaterals is made and every attempt is made to disconnect the collaterals between the portal system and the venous drainage system of the spleen. This disconnection will avoid collateralization between the portal venous and systemic venous systems.

Step 16. At this point, the operative field is checked for hemostasis.

Step 17. A generous wedge-shaped liver biopsy is taken from the left lateral segment using hemostatic chromic stitches.

Step 18. The abdominal cavity is irrigated copiously with saline, and the suture lines are inspected for hemostasis.

Step 19. The retractors are removed and the position of the NG tube is checked. The abdomen is then closed in layers using monofilament closure in two layers for the fascia closing first the posterior sheath and, subsequently, the anterior rectus sheath.

Step 20. A watertight skin closure is then applied in expectation of ascites formation. No intraperitoneal drains are placed.

Postop

The patient is then sent to the intensive care unit for careful monitoring of hemodynamics and of liver function.

Complications

Recurrent bleeding, ascites, encephalopathy.

Follow-Up

Doppler ultrasound, encephalopathy.

Chapter 89

H-Interposition Mesocaval Shunt

Michael Abecassis

Indications

The indication for an H-interposition mesocaval shunt is portal hypertensive bleeding refractory to medical and endoscopic therapy.

Preop

Hemodynamic stabilization; mesenteric venous examination (Doppler ultrasound, magnetic resonance venography).

Procedure

Step 1. The patient is positioned in the supine position under general anesthesia and endotracheal intubation. A Foley catheter is inserted for monitoring urine output and a central venous catheter is utilized for monitoring central venous pressure.

Step 2. An antiseptic prep is used from the nipple line to the groin. The body is draped in sterile fashion. Appropriate antibiotics are given prior to skin incision.

Step 3. Depending on the patient's body habitus, either a long midline or a bilateral subcostal incision can be used. In patients with narrow costal margins and a long epigastrium, the midline incision is preferred.

Step 4. The superior mesenteric vein (SMV) is exposed by lifting the transverse mesocolon and identifying the root of the small bowel mesentery. The peritoneum overlying the superior mesenteric vein is incised.

Step 5. The SMV is identified. The superior border of dissection consists of the origin of the middle colic vein. The SMV is then exposed anteriorly until it bifurcates caudally into numerous mesenteric branches. There is almost invariably a branch to the right of the SMV which needs to be ligated and divided in order to afford better mobilization of the SMV. The SMV is then mobilized circumferentially for a total distance of 2-3 cm.

Step 6. Next we turn our attention to the inferior vena cava (IVC). Exposure of the IVC requires mobilization of the third and fourth portions of the duodenum in order to sweep the duodenum upwards and create a path for the graft. Both during the dissection of the SMV and the IVC, all lymphatics should be ligated carefully in order to avoid the development of chylous/lymphatic ascites postoperatively.

Step 7. Extensive Kocherization of the second and third portions of the duodenum may be necessary in order to gain full mobilization, and this may require taking down the hepatic flexure of the colon with medial mobilization of the hepatic flexure. Once the anterior aspect of the infrahepatic IVC is exposed, minimal mobilization should ensue so that a side-biting vascular clamp can be applied for an anterior anastomosis between the IVC and the graft.

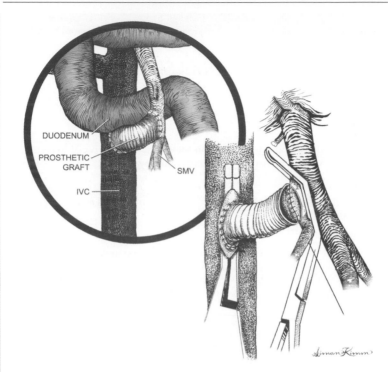

Figure 89.1. H-interposition mesocaval shunt.

Step 8. The next step is direct measurement of portal pressure. This is achieved by encircling a segment of the gastroepiploic vein easily found in the gastrocolic ligament. The proximal side is ligated and a transverse venotomy is made in the epiploic vein. A catheter is inserted in the direction of the main portal vein, and the catheter is attached to a manometer so that the portal pressure can be measured. Typically, a 5 F feeding catheter is used for this purpose, and a sterile IV tubing which is flushed with saline can be attached to the hub of the catheter and the other end passed to the anesthesiologist who can then monitor the pressure in mm Hg. This is important because a gradient is then determined between the central venous pressure and the direct portal pressure. This gradient is recorded. The catheter is secured and remains in place until completion of the shunt.

Step 9. Next, a 14 or 16 mm woven Dacron graft is used for the interposition. Alternatively, a ringed Gore-Tex of the same size can also be used. The IVC anastomosis which will be the most dependent anastomosis is performed first. This is done using fine polypropylene running suture having removed a small ellipse of anterior IVC.

Step 10. The graft is gently curved around the fourth portion of the duodenum anteromedially and is brought to the area of the previously exposed SMV. A clamp can be placed on the graft itself, removing the side-biting clamp on the IVC in order to test that suture line.

Step 11. Next, the SMV is controlled either with a side-biting vascular clamp or, alternatively, with proximal and distal control.

Step 12. A lateral venotomy is performed on the SMV and an end-to-side anastomosis is performed using fine running polypropylene suture. The SMV may be thin and one should avoid tension on this anastomosis in order to prevent tearing of the SMV. If necessary, the venotomy can be carried down caudally into one of the mesenteric branches of the SMV. Again, the superior aspect of the venotomy is dictated by the origin of the middle colic vein.

Step 13. Once the SMV to graft anastomosis is completed, the clamps are removed. The SMV should be immediately noted to be nicely decompressed and there should be a thrill palpable over the graft. There should be a gentle curve on the graft, especially as it lies inferior and slightly posterior to the fourth portion of the duodenum.

Step 14. Once hemostasis of the suture line is secured, the portal pressure is again measured directly and a second reading of the central venous pressure is recorded so that the portosystemic gradient can be calculated. A comparison of the gradient is made to the earlier recording.

Step 15. At this point, the operative field is checked for hemostasis.

Step 16. The catheter is removed from the gastroepiploic vein, and the vein is ligated.

Step 17. A generous wedge-shaped liver biopsy is taken from the left lateral segment using hemostatic chromic stitches.

Step 18. The abdominal cavity is irrigated copiously with saline and the suture lines are inspected for hemostasis.

Step 19. The retractors are removed, and the position of the NG tube is checked. The abdomen is then closed in layers using monofilament for the fascia.

Step 20. A watertight skin closure is then applied in expectation of ascites formation. No intraperitoneal drains are placed.

Postop
The patient is sent to the intensive care unit for careful monitoring of hemodynamics and of liver function.

Complications
Liver failure, thrombosis, encephalopathy.

Follow-Up
Doppler ultrasound, encephalopathy.

89

Chapter 90

Portacaval Shunts

Michael Abecassis

Indications

The indication for portacaval shunt is portal hypertensive bleeding refractory to medical and endoscopic therapy.

Preop

Hemodynamic stabilization; mesenteric venous examination (Doppler ultrasound, magnetic resonance venography).

Procedure

Step 1. The patient is positioned in the supine position under general anesthesia and endotracheal intubation. A Foley catheter is inserted for monitoring urine output and a central venous catheter is utilized for monitoring central venous pressure.

Step 2. An antiseptic prep is used from the nipple line to the groin. The body is draped in sterile fashion. Appropriate antibiotics are given prior to skin incision.

Step 3. A right subcostal incision is made. Because of the presence of portal hypertension and frequent thrombocytopenia, liberal use of cautery is advised. The incision is extended through the abdominal muscular layers, and the peritoneal cavity is entered. Ascites may be encountered which needs to be submitted for cell count and differential analysis in order to rule out the possibility of spontaneous bacterial peritonitis.

Step 4. The ligamentum teres is divided in order to facilitate access to the right upper quadrant, as is the falciform ligament. A large patent umbilical vein is often encountered within the ligamentum teres and suture ligature of this vein will prevent slippage of the ties and consequent bleeding. If necessary for exposure, the incision is extended across the midline into a left subcostal incision. Appropriate packs are placed above and behind the liver in order to bring the porta into view. A mechanical retractor is used in order to retract the costal margins proximally. A quick inspection and palpation of the hilum is essential in order to confirm patency of the portal vein which can easily be ballotted posteriorly.

Step 5. The next step is direct measurement of portal pressure. This is achieved by encircling a segment of the gastroepiploic vein easily found in the gastrocolic ligament. The proximal side is ligated and a transverse venotomy is made in the epiploic vein. A catheter is inserted in the direction of the main portal vein and the catheter is attached to a manometer so that the portal pressure can be measured. Typically, a 5 F feeding catheter is used for this purpose, and a sterile IV tubing which is flushed with saline can be attached to the hub of the catheter and the other end passed to the anesthesiologist who can then monitor the pressure in mm Hg. This is important because a gradient is then determined between the central venous

Northwestern Handbook of Surgical Procedures, edited by Richard H. Bell, Jr. and Dixon B. Kaufman. ©2005 Landes Bioscience.

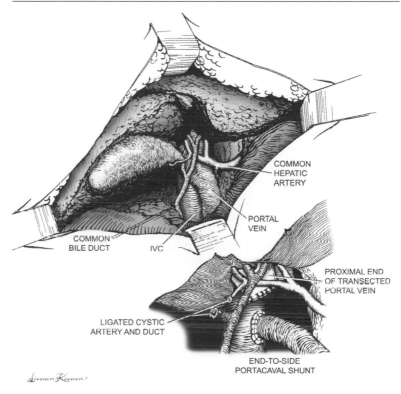

COMMON
HEPATIC
ARTERY

PORTAL
VEIN

COMMON
BILE DUCT   IVC

PROXIMAL END
OF TRANSECTED
PORTAL VEIN

LIGATED CYSTIC
ARTERY AND DUCT

END-TO-SIDE
PORTACAVAL SHUNT

Figure 90.1. Portacaval shunt.

pressure and the direct portal pressure. This gradient is recorded. The catheter is secured and remains in place after completion of the shunt. Therefore, it is important that the catheter not be placed past the location of portal vein transection.

Step 6. The next step involves mobilizing the portal vein. It may be necessary to perform a cholecystectomy in order to achieve complete exposure of the posterior hilum. The cystic structures are transected and the gallbladder is dissected away from the fossa taking great care not to violate Glisson's capsule.

Step 7. The hepatic arterial anatomy is determined at this point, and aberrant right hepatic arterial anomalies are identified. An aberrant right hepatic artery emanating either from the superior mesenteric artery or directly from the aorta can be found posterior to the bile duct and lateral to the portal vein and may require extensive mobilization prior to mobilization of the portal vein. The portal vein can be approached anteriorly by encircling the common bile duct, placing a vessel loop around it and retracting laterally on the vessel loop while identifying the main hepatic artery, encircling the artery with a second vessel loop, and retracting medially on the artery. The plane between the hepatic artery and the bile duct will lead directly to the anterior aspect of the portal vein and is typically devoid of important vascular or biliary structures. Typically, the superior dissection of this anterior approach is limited by the right hepatic artery, and the inferior limit of the dissection is encompassed by the gastroduodenal artery.

Step 8. Division of the gastroduodenal artery may be necessary to better expose the portal vein. Great care should be taken not to devascularize the bile duct, and all lymphatic tissue in the area should be carefully ligated in order to prevent the development of postoperative chylous ascites. Once the bile duct is mobilized, a vein retractor can be very useful in retracting the bile duct anteriorly so that the portal vein can be then be approached posterolaterally.

Step 9. Using blunt dissection with a peanut, the portal vein is mobilized posterolaterally from the rest of the porta. This dissection is done under direct vision, and occasionally small branches of the portal vein may need to be divided. This dissection is carried out until the bifurcation of the left and right portal veins is visualized. It is important to appreciate that most of this dissection is performed from a posterior and lateral approach to the porta hepatis. Once the portal vein can be encircled with a vessel loop, the dissection proceeds more easily with retraction of the portal vein posterolaterally away from the hilum.

Step 10. The next step consists of identifying the anterior aspect of the infrahepatic inferior vena cava (IVC). The peritoneum between the caudate lobe and the IVC is incised in order to expose the IVC anteriorly. Caudate venous branches may need to be divided to improve exposure. The cava is dissected free of the peritoneum caudally and needs to be mobilized enough for a side-biting vascular clamp to be applied. The direction of the shunt needs to be envisioned in order to plan the end-to-side shunt so that no kinking of the portal vein occurs. This may require dissection posterior to the portal vein, typically of lymphatic tissue and, again, these should be ligated carefully in order to avoid postoperative chylous/lymphatic ascites. In contrast to a side-to-side portacaval shunt, minimal dissection of the cava should be necessary.

Step 11. At this point, a vascular clamp is applied to the portal vein proximally, and the liver side of the portal vein should be controlled. Preferably, the left and right portal veins are tied individually with heavy silk and the portal vein is transected short of the bifurcation in order to prevent slippage of the portal venous ties. Alternatively, the liver side of the portal vein can be clamped with a crushing clamp and oversewn with fine polypropylene. Finally, vascular staples may be applied to the hepatic side of the portal vein. Regardless, the vein is transected and swung posterolaterally to the IVC.

Step 12. A side-biting vascular clamp is then applied to the IVC, and optimally an ellipse of vein is removed from the anterior aspect of the IVC, leaving a defect that matches the orifice of the portal vein.

Step 13. An end-to-side anastomosis is constructed, performing the back wall anastomosis from the inside and completing the anterior wall anastomosis. There should not be much redundancy of the portal vein in order to prevent kinking, and great care should be taken so that the portal vein is not angled acutely possibly resulting in obstruction of the shunt.

Step 14. The side-biting clamp on the cava is released first so that any bleeders in the anastomotic line can be identified, and the portal vein clamp is finally released. There should be a thrill palpable over the shunt. The portal vein should be soft and noticeably decompressed.

90

Step 15. Once hemostasis of the suture line is secured, the portal pressure is again measured directly and a second reading of the central venous pressure is recorded so that the portosystemic gradient can be calculated. A comparison of the gradient is made to the earlier recording.

Step 16. At this point, the operative field is checked for hemostasis. The bile duct and hepatic artery are inspected. The catheter is removed from the gastroepiploic vein and the vein is ligated.

Step 17. A generous wedge-shaped liver biopsy is taken from the left lateral segment using hemostatic chromic stitches.

Step 18. The abdominal cavity is irrigated copiously with saline, and the suture line on the liver side of the portal vein is inspected for hemostasis.

Step 19. The retractors are removed, and the position of the NG tube is checked. The abdomen is then closed in layers using monofilament closure in two layers for the fascia closing first the posterior sheath and, subsequently, the anterior rectus sheath.

Step 20. A watertight skin closure is then applied in expectation of ascites formation. No intraperitoneal drains are placed.

Postop

The patient is sent to the intensive care unit for careful monitoring of hemodynamics and of liver function.

Complications

Ascites, liver failure, thrombosis, encephalopathy.

Follow-Up

Doppler ultrasound, encephalopathy.

Chapter 91

Liver Transplantation

Jonathan Fryer

Indications

Liver transplantation is indicated in circumstances where a life-threatening pathological process that involves the liver cannot be overcome, without replacing the entire liver. Most commonly, this occurs when end-stage liver disease (ESLD) has developed in a cirrhotic liver. Other situations where liver transplantation may be needed when cirrhosis is not present include acute liver failure, nonmetastatic tumors that are otherwise unresectable, polycystic liver disease, severe hepatic trauma, or acute failure of a transplanted liver. Disease entities that commonly cause cirrhosis in adults include hepatitis B and C, alcohol abuse, nonalcoholic steatohepatitis (NASH), primary biliary cirrhosis (PBC), and primary sclerosing cholangitis (PSC). In children common causes of cirrhosis include biliary atresia and metabolic liver diseases like Wilson's or alpha 1 antitrypsin deficiency. Other conditions, like acute liver failure and inborn errors of metabolism, may also necessitate transplantation in children. Not all cirrhotic patients need liver transplants. Some cirrhotic patients who have not developed significant complications attributable to their liver disease (i.e., ESLD) may not require liver transplant. The complications of cirrhosis that typically warrant consideration of transplantation include poorly controlled portal hypertension, bleeding, ascites or peripheral edema, hepatic encephalopathy, hydrothorax, hepatorenal syndrome, hepatopulmonary syndrome, hepatocellular carcinoma, severe fatigue, and severe pruritus. If these conditions cannot be adequately controlled with more conservative therapies, liver transplantation should be considered.

Preop

Since one cannot always predict when cadaveric donor organs will become available, liver transplant candidates on the waiting list require ongoing monitoring to ensure they are kept in optimal condition for surgery at all times. They must be followed closely for infection, bleeding, malnutrition, or other problems that could compromise their transplant eligibility when a liver donor becomes available. If they have hepatocellular carcinoma, they must be frequently reevaluated to rule out tumor progression. Conversely, if a segment of liver from a living donor is going to be used for transplantation, a single preoperative evaluation of the donor and recipient is sufficient since the transplant can be scheduled electively. When a suitable cadaveric donor liver has been offered, the donor procurement team is sent to the donor hospital where they perform a final inspection of the liver to ensure that it is a suitable size match for the recipient and does not have significant traumatic injuries or fatty changes. A liver biopsy is sometimes necessary to rule out a fatty liver or other pathologic entities. Meanwhile, the potential recipient is brought in to the hospital and prepared for surgery. If final evaluations of the donor liver and the liver

Northwestern Handbook of Surgical Procedures, edited by Richard H. Bell, Jr. and Dixon B. Kaufman. ©2005 Landes Bioscience.

transplant candidate reveal no surprises they are cross-matched for packed red blood cells, plasma, and platelets and transferred to the operating room.

After intubation and induction of general anesthesia, several large-bore venous lines are placed to provide hemodynamic monitoring and to allow rapid infusion of blood products, medications, and fluids. Arterial lines are placed both for blood pressure monitoring and for obtaining blood specimens to monitor arterial blood gases, electrolytes, CBC, and coagulation. A Foley catheter is placed to monitor urine output. To provide much-needed gastric decompression, an orogastric or nasogastric tube is placed at the beginning of the case, with great care given not to initiate bleeding from esophagogastric varices. In addition to the chest and abdomen, the left groin and left upper arm, including axilla, are shaved and prepped to provide sterile access to the sapheno-femoral junction and axillary veins respectively, should cannulation for veno-veno bypass be necessary.

Procedure

Step 1. The patient is positioned supine with both arms partially extended (45°) to optimize access to the axillae for veno-veno bypass without risking hyperextension injuries. The legs are partially abducted and elevated to optimize exposure to the groins for veno-veno bypass and to promote venous return.

Step 2. The entire abdomen, chest, left axilla, and left groin should be surgically prepped and draped in continuity. Access to the chest may be necessary intraoperatively to optimize surgical exposure, insert chest tubes, or to administer manual cardiac compression in the event of cardiac arrest. Access to the left groin and axilla will be needed if veno-veno bypass becomes necessary. The right groin and/or axilla can also be used if necessary.

Step 3. A bilateral subcostal incision is used often with a cephalad midline extension to the xiphoid process. To provide exposure to the right lobe the right extension of the subcostal incision usually exceeds the left.

Step 4. The umbilical vein, encountered in the midline, is encircled, ligated, and divided. The falciform ligament is divided using cautery as far cephalad as exposure allows.

Step 5. Using a fixed mechanical retractor, the rib cage is retracted cephalad and lateral to optimize exposure to suprahepatic structures. Similarly the infrahepatic viscera are gently retracted caudad to optimize exposure to the porta hepatis and infrahepatic structures. Adhesions should be carefully divided prior to placing retractors to avoid tearing adjacent structures.

Step 6. With retractors in place, the falciform ligament is divided cephalad to where it splits into left and right coronary ligaments. Taking care to protect the stomach and spleen, the left triangular and coronary ligaments are divided from lateral to medial using cautery.

Step 7. Retracting the left lobe of the liver to the right, the gastrohepatic ligament (lesser omentum) is exposed. If a replaced or accessory left hepatic artery is identified, it should be ligated and divided. The gastohepatic ligament is divided from the cut edge of the left coronary ligament to the porta hepatis using cautery, thereby exposing the caudate lobe.

Step 8. The porta hepatis is dissected in a plane close to the liver. All venous and lymphatic structures should be ligated and divided. The hepatic arteries and common hepatic duct are ligated and divided close to the liver. The cystic duct and artery are ligated and divided close to the gallbladder. The portal vein is skeletonized

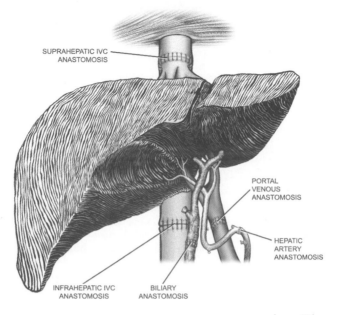

Figure 91.1. Liver transplantation. Caval anastomoses.

along an adequate length to allow placement of vascular clamps and a bypass cannula if portal decompression is necessary using veno-veno bypass.

Step 9. The right lobe of the liver is gently retracted to the left exposing the right triangular ligament which is divided using cautery. The peritoneal attachments to the right lobe are divided as gradually progressive leftward retraction is applied until the retrohepatic vena cava is exposed.

Step 10. The left and caudate lobes are retracted to the right exposing the posterior peritoneal reflections of the caudate and retrohepatic vena cava, which are carefully divided using cautery. Care should be taken not to injure venous branches of the cava.

Step 11. There are two techniques commonly utilized for performing the caval anastomoses: the standard technique (Fig. 91.1), and the cavaplasty technique (Fig. 91.3). With both techniques the suprahepatic cava is encircled. With cavaplasty, no further dissection is performed. With the standard technique, the infrahepatic vena cava is also encircled. The right adrenal vein should be identified to avoid injury, and divided if necessary. The posterior surface of the retrohepatic vena cava and the caudate lobe are freed from their retroperitoneal attachments.

Step 12. In coordination with anesthesia, the portal vein, infrahepatic cava, and suprahepatic cava are sequentially cross-clamped. If the patient tolerates this hemodynamically, the liver is excised. The portal vein is divided close to the liver. With the standard technique the entire retrohepatic cava is removed with the liver while with the cavaplasty technique, the liver is separated from the cava which is kept intact. The explanted liver is sent to pathology.

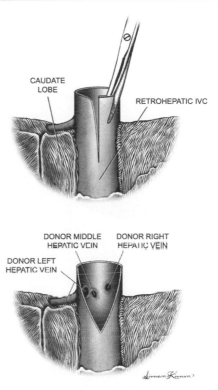

CAUDATE LOBE

RETROHEPATIC IVC

DONOR MIDDLE HEPATIC VEIN

DONOR RIGHT HEPATIC VEIN

DONOR LEFT HEPATIC VEIN

91

Figure 91.2. Liver transplantation. Caval anastomoses by the cavaplasty technique.

Step 13. The donor liver is brought up to the surgical field. The recipient is placed in reverse Trendelenburg to optimize exposure. With the standard technique, the suprahepatic cavae of the donor and recipient are anastomosed end-to-end with 4-0 polypropylene in a running fashion. With the cavaplasty technique, a posterior slit is made from the cephalad end of the donor retrohepatic cava (Fig. 91.2) and an anterior slit is made in the recipient cava extending caudad from the joined hepatic vein orifices. A triangular anastomosis is then performed between the posterior aspect of the donor cava and the anterior aspect of the recipient cava using 4-0 polypropylene in a running fashion. (Fig. 91.3).

Step 14. The patient is placed in slight Trendelenburg. To eliminate residual preservation solution, the liver graft is flushed with 1L of Ringer's Lactate which is infused via the portal vein and vented via the infrahepatic vena cava. With the cavaplasty technique the infrahepatic cava is then stapled closed. With the standard technique, the infrahepatic cavae of the donor and recipient are anastomosed end-to-end using 4-0 polypropylene (Fig. 91.1).

Step 15. The donor and recipient portal veins are cut to appropriate lengths to allow end-to-end anastomosis without tension or redundancy. The anastomosis is performed using 5-0 polypropylene and a running technique. A growth stitch (i.e.,

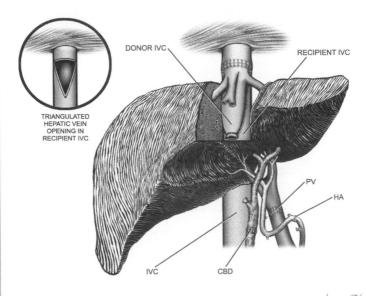

DONOR IVC

RECIPIENT IVC

TRIANGULATED
HEPATIC VEIN
OPENING IN
RECIPIENT IVC

PV

HA

IVC   CBD

Figure 91.3. Liver transplantation. Caval anastomoses with a triangulated hepatic vein opening in the recipient IVC.

"air-knot") is left to allow expansion at the anastomotic site. The portal, infrahepatic, and suprahepatic caval clamps are sequentially removed, thereby reperfusing the liver.

Step 16. After surgical hemostasis has been achieved, the arterial anastomosis is performed. After being flushed with heparinized saline, small atraumatic vascular clamps are applied to the donor and recipient arteries. The sites for anastomosis with the donor and recipient arteries are selected based on their relative calibers, lengths, and quality. A tensionless anastomosis without excessive redundancy between two vessels of comparable quality and luminal diameter is the goal. A common practice is to fashion a Carrel patch with the donor artery using the origin of the celiac trunk, or "branch patches" in donor and recipient vessels using any suitable arterial bifurcation, to help prevent narrowing at the anastomotic site. The anastomosis is performed using 6-0 polypropylene and a running technique. When complete the clamps are removed. If the recipient artery is thought to be unsuitable, an arterial conduit, fashioned from donor arteries, can be anastomosed to the infrarenal aorta and passed through a retrocolic or retropancreatic tunnel where it is anastomosed to the donor artery.

Step 17. The biliary anastomosis is then performed. Most commonly, this is performed end-to-end between donor and recipient ducts using an 5-0 absorbable monofilament suture and an interrupted technique. Great care must be taken to preserve the blood supply to both the donor and recipient ducts. T-tubes or stents are unneccesary. If a duct-to-duct anastomosis cannot be achieved safely due to recipient disease (i.e., PSC, biliary atresia) or donor issues (i.e., too small), a biliary enteric anastomosis is performed using a loop of jejunum fashioned as a Roux en Y. This anastomosis is similarly performed using interrupted 5-0 absorbable monofilament sutures. **Step 18.** Prior to closure a vigorous attempt should be made to eliminate all surgical bleeding and reverse coagulopathy. When this is acheived the retractors are removed. Consideration should be given to placing Jackson-Pratt drains to monitor bleeding and detect bile leaks. Closure should be performed in two layers using 0-polypropylene sutures.

Postop

Because bleeding, sepsis, low systemic vascular resistance, and massive fluid shifts are common in the immediate posttransplant period, close monitoring in an ICU setting is essential until hemodynamic stability is achieved. However, when stable, patients should be extubated and transferred to the regular surgical floor. With most straightforward transplants the recipients can be transferred out of ICU within 24 48 hours. As soon as feasible, patients should be mobilized and nasogastric (NG) tubes, Foley catheters, and invasive vascular lines removed and oral intake initiated. In patients with Roux-en-Y anastomoses, the NG tube should be retained until intestinal function resumes. In optimal circumstances these patients can be discharged in 2-3 days. In patients that experience complications or that were severely debilitated pretransplant, a longer hospital stay is sometimes necessary.

Complications

Potential postoperative complications include bleeding, infection, bile duct problems, hepatic arterial thrombosis, and fluid retention.

Follow-Up

In the first several postoperative days, close monitoring of liver enzymes (ALT, AST, alkaline phosphatase) and parameters of liver function (INR, bilirubin, acidosis) is necessary to evaluate for evidence of hepatic artery thrombosis (HAT) or primary nonfunction (PNF) of the liver graft. If hepatic artery patency is in question, liver ultrasound with Doppler evaluation of the hepatic vasculature and/or hepatic angiography are required. If HAT or PNF are diagnosed within the first 7 days posttransplant, a recipient is eligible for relisting for another liver transplant as a status 1. Once stability is achieved, liver transplant recipients can be discharged home as soon as they are mobile and capable of taking care of their own bodily needs. In the first several months posttransplant, recipients need to be closely monitored for rejection, infectious complications, biliary complications, recurrent disease, and complications related to immunosuppressive drugs (nephrotoxicity, neurotoxicity, neutropenia, hyperglycemia, gastrointestinal toxicity) or to generalized immunosuppression such as posttransplant lymphoproliferative disorders. In the long-term, the most significant concerns are recurrent disease (hepatitis C, hepatitis B, alcohol abuse), chronic rejection, and the consequences of long-term immunosuppressive therapy such as renal failure, hypertension, diabetes, and bone disease. Hepatitis C is the most common etiology leading to liver transplantation and these patients are especially difficult to follow posttransplant. Hepatitis C recurs in essentially all liver transplant recipients. While hepatitis C recurrence is usually indolent and slowly progressive, approximately 10% of hepatitis C recipients experience an aggressive recurrence rapidly progressing to end-stage liver disease within months. Further, hepatitis C is difficult to differentiate histologically from other hepatic inflammatory processes, including acute rejection.

Chapter 92

Pancreas Transplantation

Dixon B. Kaufman

Indications

Pancreas transplantation is indicated for Type I diabetes complicated by difficulty with conventional exogenous insulin therapy. Often patients with chronic or end-stage renal failure with diabetes are candidates for combined pancreas and kidney transplantation.

Preop

General orotracheal anesthesia, central venous catheter, Foley bladder catheter, and arterial line (optional).

Procedure

Step 1. Prone position. Midline abdominal incision starting approximately 6-8 cm below xiphoid and extending to the pubis.

Step 2. Mobilize the ascending colon medially by taking down the avascular plane at white line of Toldt to the hepatic flexure to expose the common iliac vein and artery. Alternatively, open the retroperitoneal tissue medial to the ascending colon to expose the common iliac artery, distal vena cava, and common iliac vein.

Step 3. Partially mobilize and expose the distal vena cava and common iliac vein by suture ligation and division of overlying lymphatics.

Step 4. Place vascular clamps on the proximal common iliac vein.

Step 5. Anastomose the pancreas allograft portal vein (with or without short extension graft from donor external iliac vein) using 5-0 monofilament suture.

Step 6. Place vascular clamps (e.g., Fogarty clamps with inserts) on the common iliac artery (or external iliac artery) distally relative to the position of venous anastomosis.

Step 7. Anastomose the common iliac artery portion of the pancreas allograft extension Y-graft that was previously anastomosed to the donor splenic and superior mesenteric arteries (from donor internal/external/common iliac artery complex) to the recipient common iliac artery slightly distal to the position of the venous anastomosis using 6-0 monofilament suture.

Step 8. Slowly remove venous clamps and then release the arterial clamps to allow full perfusion of the pancreas allograft.

Step 9. Control bleeding points with suture ligation or electrocautery.

Step 10. Perform donor-to-recipient duodenoenterostomy to midjejunum approximately 20-30 cm distal to ligament of Treitz.

Step 11. Handsewn two-layer anastomosis performed with bowel closed by first placing back wall of 3-0 silk sutures as interrupted Lembert stitches approximating the antimesenteric borders of the donor duodenum to the recipient jejunum. Length of anastomosis approximately 3 inches.

Northwestern Handbook of Surgical Procedures, edited by Richard H. Bell, Jr.
and Dixon B. Kaufman. ©2005 Landes Bioscience.

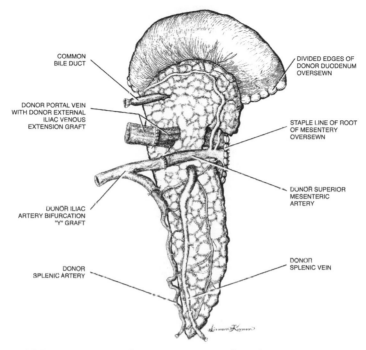

COMMON
BILE DUCT

DIVIDED EDGES OF
DONOR DUODENUM
OVERSEWN

DONOR PORTAL VEIN
WITH DONOR EXTERNAL
ILIAC VENOUS
EXTENSION GRAFT

STAPLE LINE OF ROOT
OF MESENTERY
OVERSEWN

DONOR SUPERIOR
MESENTERIC
ARTERY

DONOR ILIAC
ARTERY BIFURCATION
"Y" GRAFT

DONOR
SPLENIC VEIN

DONOR
SPLENIC ARTERY

92

Figure 92.1. Pancreas transplantation. Pancreas allograft.

Step 12. Open donor duodenum, aspirating contents, and open adjacent recipient jejunum.

Step 13. Place inner layer mucosal-to-mucosal anastomotic suture line using 3-0 absorbable monofilament suture as a running hemostatic stitch.

Step 14. Place outer second anastomotic layer of interrupted Lembert stitches using 3-0 silk suture.

Step 15. Position the pancreas allograft with tail tucked into the retrocystic pouch of Douglas and the pancreaticoduodenal head cephalad.

Step 16. Irrigate abdomen with 2 l of antibiotic solution and close midline fascia in a single layer with running 1-0 monofilament suture.

Postop

Continue immunosuppressive medical therapy according to institutional protocols. Administer mild anticoagulation and/or antiplatelet agents to minimize graft thrombosis risk. Administer parenteral antibiotics for 3-7 days to reduce infectious risk. Follow hemoglobin levels for bleeding. Follow serial blood glucose measurements to assess pancreatic function.

Complications

Rejection, vascular thrombosis, transplant pancreatitis, enteric leak and intraabdominal abscess, GI bleeding from duodenoenteric anastomosis.

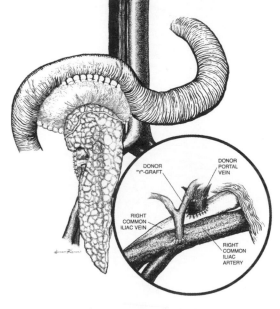

92 Figure 92.2. Pancreas transplantation. Vascular anastomoses.

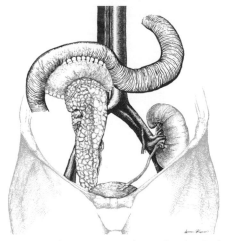

Figure 92.3. Pancreas transplantation. Complete with tail tucked into the retrocystic pouch of Douglas and the pancreaticoduodenal head cephalad.

Follow-Up

Immunosuppression to prevent rejection. Monitoring of chemistries to assess pancreas transplant function and to ensure adequate exposure to immunosuppression.

Section 7: Vascular Surgery
Section Editor: William H. Pearce

Chapter 93

Carotid Endarterectomy

Mark D. Morasch

Indications

Carotid endarterectomy is indicated for prevention of stroke in patients with symptomatic carotid bifurcation stenosis (greater than 50%) and asymptomatic carotid bifurcation stenosis (greater than 60%) contingent upon the patient's estimated perioperative stroke and death rate.

Preop

General oral endotracheal anesthesia or local regional anesthesia. Continuous radial arterial blood pressure monitoring.

Procedure

Step 1. Position the patient supine with the torso and head slightly elevated and turned to expose the neck on the surgical side.

Step 2. Longitudinal incision along the sternocleidomastoid muscle in the upper neck or transverse incision in a skin crease with the creation of subplatysmal flaps.

Step 3. Mobilize the sternocleidomastoid muscle posteriorly to expose the jugular vein, taking care not to injure the greater auricular nerve.

Step 4. Lateral mobilization of the jugular vein and jugulodigastric lymph node group. Ligate and divide all venous branches from the anterior aspect of the jugular vein. This includes ligation and division of the facial vein in most cases. The anterior aspect of the jugular vein should be completely mobilized from the level of the strap muscles to a site above the angle of the mandible.

Step 5. Gentle circumferential dissection of the common carotid artery at the base of the neck. Care must be taken to avoid vigorous manipulation of the common carotid artery. Similarly, the vagus nerve should be retracted laterally with minimal dissection or manipulation.

Step 6. Circumferential dissection of both the distal internal carotid artery and the external carotid artery, again, taking care to avoid excessive manipulation of the vessels. The carotid bulb should be left undissected until after clamping. Both the vagus and hypoglossal nerves require gentle retraction away from the operative field.

Step 7. Administration of systemic heparin for anticoagulation through an intravenous catheter; measurement of carotid back stump pressures (optional); and clamping of internal, external, and common carotid arteries, in that order.

Step 8. Mobilization of the area of the carotid bifurcation (after clamps are applied) to allow rotation of the carotid bifurcation. This allows for creation of a longitudinal arteriotomy along the lateral aspect of the common and internal carotid

Northwestern Handbook of Surgical Procedures, edited by Richard H. Bell, Jr. and Dixon B. Kaufman. ©2005 Landes Bioscience.

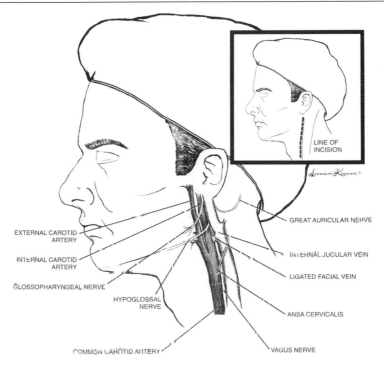

EXTERNAL CAROTID ARTERY

INTERNAL CAROTID ARTERY

GLOSSOPHARYNGEAL NERVE

HYPOGLOSSAL NERVE

COMMON CAROTID ARTERY

LINE OF INCISION

GREAT AURICULAR NERVE

INTERNAL JUGULAR VEIN

LIGATED FACIAL VEIN

ANSA CERVICALIS

VAGUS NERVE

93

Figure 93.1. Carotid endarterectomy.

arteries through the area of bifurcation stenosis. Eversion endarterectomy can be performed as an alternative technique. Eversion endarterectomy is completed after transversely amputating the internal carotid artery just distal to the bifurcation.

Step 9. The endarterectomy is started by identifying the appropriate plane between the inner one-third and the outer two-thirds of the media using an endarterectomy spatula. The plaque specimen is then transected in the common carotid artery proximal to the bulk of the disease. Eversion technique is utilized to clear the external carotid artery of plaque. Great care must be taken to feather the distal end of the plaque as it ends in the internal carotid artery. Gentle downward traction on the plaque by the operator or by the assistant will facilitate the creation of a smooth endpoint.

Step 10. Irrigate the remaining vessel wall to identify and remove remaining small fronds of media and to create a smooth new vessel lumen.

Step 11. Close the longitudinal arteriotomy with a single small monofilament suture. Alternatively, a patch closure can be utilized to further widen the bulb if necessary. Polytetrafluoroethylene (PTFE), Dacron, or thigh saphenous vein are the usual choices for patch closure.

Step 12. After backbleeding and forebleeding all three vessels and vigorously flushing the site with heparinized saline, the arteriotomy can be closed completely and flow reestablished. The clamp on the internal carotid should be released last.

Step 13. Completion imaging of the endarterectomy site utilizing angiography, intraoperative duplex ultrasound, or angioscopy.

Step 14. Hemostasis and wound closure by reapproximating the platysma and then the skin with a plastics-type closure.

Step 15. A carotid shunt can be placed routinely or selectively in patients considered high risk for perioperative stroke.

Step 16. A small, closed-suction drain can be left in place at the discretion of the surgeon.

Postop

The patient should be monitored in the recovery room for at least 2 hours and, if neurologically and hemodynamically stable, may be transferred postoperatively to a surgical ward. If complications arise or if the patient requires intravenous infusions to control hypertension, hypotension, or arrhythmias, the patient should be transferred from the recovery room to an intensive care unit for further monitoring.

Complications

Permanent or transient central neurologic deficits (stroke or TIA); bleeding complications requiring neck reexploration and hematoma evacuation; peripheral nerve injuries including injuries to the hypoglossal, vagus, recurrent laryngeal, marginal mandibular, posterior auricular, and, more rarely, the glossal pharyngeal and spinal accessory nerves.

Follow-Up

Patients are usually discharged on the day following the operation and should be seen back in the outpatient setting in 3 weeks to assure normal wound healing. The patient should be monitored with duplex ultrasound after 6 months and yearly thereafter.

Chapter 94

Repair Infrarenal Aortic Aneurysm: Elective

Jon S. Matsumura

Indications

Elective repair of an infrarenal aortic aneurysm is indicated for an abdominal aortic aneurysm that is 5.5 cm or larger, enlarging >5 mm/6 months, symptomatic, or associated with other risk factors for rupture.

Operative Principles

Secure vascular control and visualization of the aorta and iliac arteries usually requires generous exposure and lighting. Specific attention is given to avoiding inadvertent venous injuries. Gentle handling and retraction of the bowel and keeping the small intestine inside the abdomen can minimize postoperative ileus and shorten recovery. Both mental preplanning and manual dexterity are contributory elements for rapid vascular reconstruction which minimizes cross-clamp time, bleeding, and occlusion risk. Meticulous review of the preoperative imaging assists in formulation of this individualized cognitive plan.

Preop

The patient is given a mild, nondehydrating bowel laxative the day before. One gram of cefazolin is given intravenously. Two units of packed red blood cells are crossmatched. A cell saver collection is attached to the suction but is only processed if there is a need for transfusion. The abdomen and groins are shaved.

Procedure

Step 1. Supine position, bladder catheter, prep from nipples to the knees, midline incision from the xiphoid to lower abdomen.

Step 2. Avoid bowel injury when entering the peritoneum; anterior adhesions are taken down carefully. Any enterotomy would necessitate aborting the aortic reconstruction.

Step 3. Retract the transverse colon cranially, small bowel to the right, and sigmoid colon to the left using self-retaining retractor system.

Step 4. Proximal aortic neck exposure: mobilize the duodenum, identify and loop the left renal vein, and finger dissect tissue lateral to infrarenal neck. Test vascular clamp application.

Step 5. Distal exposure of iliac arteries: preoperative imaging should predict the extent of dissection but reassess after direct observation and palpation of the arteries, as a different graft configuration may be indicated. If necessary, expose femoral artery for anastomosis. When these dissections are mandatory, they can be performed prior to opening the abdomen to decrease the laparotomy time. Create retroureteral/retrocolic tunnels by tunneling just anterior to the external and common iliac arteries.

Northwestern Handbook of Surgical Procedures, edited by Richard H. Bell, Jr. and Dixon B. Kaufman. ©2005 Landes Bioscience.

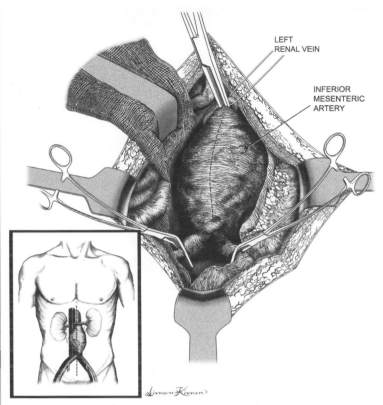

Figure 94.1. Repair of infrarenal aortic aneurysm. Elective.

Step 6. Systemic heparin may be given if the patient has minimal bleeding.

Step 7. Open aneurysm sac and oversew lumbar arteries with 2-0 braided nonabsorbable suture. The inferior mesenteric artery may backbleed from the underside of the left edge of the sac and needs to be looked for specifically.

Step 8. Select a graft, trim the aortic portion if a bifurcated prosthesis, and suture to the aorta with 3-0 polypropylene. Flush anastomosis, move clamp onto graft, and check hemostasis (particularly posterior half) prior to working on distal. Complete distal anastomosis and flush graft.

Step 9. Verify hemostasis of all suture lines and recheck lumbar vessels (after distal flow is restored, back pressure may increase.) Observe backbleeding through inferior mesenteric artery and reimplant if bleeding is sluggish or both hypogastric arteries are occluded. Palpate for femoral pulses.

Step 10. Close the sac and approximate retroperitoneal tissues with running 2-0 absorbable suture so that the aortic graft is covered and not in contact with bowel.

Step 11. Close the abdomen with running #2 nylon with the knots buried. Avoid electrocautery near this suture line.

Step 12. Check distal pulses/Doppler indices prior to moving patient off the operating room table.

Postop

Rapid transport to intensive/postanesthetic care unit is necessary to minimize the gap in monitoring, resuscitation, and rewarming. Monitor organ function, particularly myocardial and renal function. Correct coagulation and electrolyte abnormalities. Maintain treatment of chronic medical comorbidities. Prescribe appropriate pain control. Check the chest X-ray if central line attempted.

Complications

Clinical outcome after technically successful aortic operations is dependent on anticipation, attempted prevention, early identification, and appropriate management of the many potential complications. Patients with aortic aneurysm often have serious pathology of critical organs that remains uncorrected after the aortic operation. Bleeding, renal failure, myocardial infarction, and limb ischemia are the main concerns in the first 24 postoperative hours. In the first few days, infections (pulmonary, urinary, and line), colon ischemia, fluid overload, and electrolyte/acid-base imbalance are additional concerns. After several days, wound complications, gastrointestinal hemorrhage, nutritional depletion, deconditioning, mental status changes, venous thromboembolism, and pressure ulcers come into play. While many of these complications become clinically manifest later, earlier attention can arrest the problems at a subclinical level. Long-term complications include ventral hernia, prosthetic graft infection, aortoenteric fistula, graft limb occlusion, and subsequent development of aneurysm.

Follow-Up

Patients are seen frequently postoperatively to monitor the above complications. After recovery, they are seen annually for clinical exam and to reinforce lifestyle modification. A CT scan is done every 5 years to detect subsequent aneurysms.

94

Chapter 95

Repair Infrarenal Aortic Aneurysm: Emergent for Rupture

Jon S. Matsumura

Indications

The indication for emergent repair of infrarenal aortic aneurysm is rupture.

Operative Principles

Rapid vascular control of the aorta requires timely diagnosis and decisive tranportation to the operating room. Delayed resuscitation until bleeding is controlled decreases dilution-related coagulopathy. Prelaparotomy transbrachial aortic occlusion balloon under local anesthetic may be feasible in some centers.

Preop

The patient is not sedated or paralyzed until just prior to laparotomy to avoid complete vascular collapse. Two grams of cefazolin are given intravenously. Type and cross 6 units packed red blood cells and order cell saver. Shave abdomen and groins.

Procedure

Step 1. Supine position, bladder catheter, prep from nipples to the knees, midline incision, avoid bowel injury.

Step 2. Supraceliac aortic control: The left lobe of the liver is retracted to the patient's right and the stomach to the left. Incise the gastrohepatic ligament to enter the lesser sac. Use the nasogastric (NG) tube to palpate the esophagus and move it to the left to expose the right crus of the diaphragm. Either use the aortic compressor to pinch aorta against the spine or divide the crus, finger dissect the lateral sides of the aorta, and then place the vascular clamp over the index and middle fingers onto the aorta.

Step 3. Retract the transverse colon cranially, small bowel to the right, and sigmoid colon to the left using self-retaining retractor system or two second assistants.

Step 4. The proximal neck is often dissected by hematoma, otherwise mobilize the duodenum, identify and loop the left renal vein, and finger dissect tissue lateral to the infrarenal neck. Once a vascular clamp is applied here, remove the supraceliac clamp. If the patient has no blood pressure, a rapid alternative to supraceliac control is to reach into the retroperitoneal hematoma and feel for the aneurysm. Move the hand cranially until the proximal neck is palpated and pinch the neck between the thumb and fingers until the area is exposed and clamped.

Step 5. Distal control of the iliac arteries obtained. Small iliac aneurysms may not be addressed in emergent operation for rupture, thus avoiding the extra time for a

Northwestern Handbook of Surgical Procedures, edited by Richard H. Bell, Jr. and Dixon B. Kaufman. ©2005 Landes Bioscience.

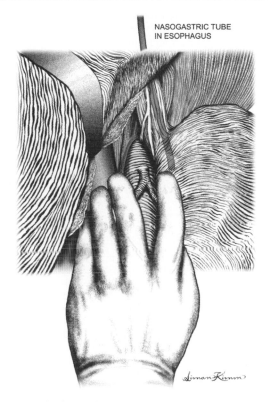

NASOGASTRIC TUBE
IN ESOPHAGUS

95

Figure 95.1. Repair of infrarenal aortic aneurysm. Emergent. Use the NG tube to palpate the esophagus and move it to the left to expose the right crus of the diaphragm.

bifurcated graft or femoral dissections. If necessary, expose the femoral arteries—beware of unsuspected femoral aneurysms—and create retroureteral/retrocolic tunnels.

Step 6. Heparin may be given if the patient has minimal bleeding and blood loss has been limited.

Step 7. Open the sac and oversew the lumbar arteries with 2-0 braided non-absorbable suture. The inferior mesenteric artery may backbleed from the underside of the left edge of the sac and needs to be looked for specifically.

Step 8. Select a graft, trim the aortic portion if a bifurcated prosthesis, and suture to the aorta with 3-0 polypropylene. Flush the anastomosis, move clamp onto graft, and check hemostasis (particularly posterior half) prior to working on the distal end. Complete the distal anastomosis and flush the graft.

Step 9. Verify hemostasis of all suture lines and recheck the lumbar vessels. After distal flow is restored, back pressure may increase. Observe backbleeding through the inferior mesenteric artery and reimplant it if bleeding is sluggish or both hypogastric arteries are occluded. Palpate for femoral pulses.

Step 10. Close the sac and approximate retroperitoneal tissues with running 2-0 absorbable suture so that the aortic graft is covered and not in contact with the bowel.

Step 11. Close the abdomen with running #2 nylon. Consider retention sutures if fascia attenuated or leave the abdomen open if abdominal compartment syndrome likely.

Step 12. Check distal pulses/Doppler indices prior to moving the patient off the operating room table.

Postop

Rapid transport to the intensive/postanesthetic care unit is necessary to minimize the gap in monitoring, resuscitation, and rewarming. Third space losses, ongoing bleeding, and reperfusion syndromes generate significant intravascular fluid deficits. Monitor organ function, particularly myocardial and renal function. Correct coagulation and electrolyte abnormalities. Maintain treatment of chronic medical comorbidities. Often complete ventilator support and continuous sedative drips are necessary in the early postoperative period. Check all tube placements and consider relocating sites, as initial insertion is often done in extreme circumstances.

Complications

Clinical outcome after technically successful aortic operations is dependent on anticipation, attempted prevention, early identification, and appropriate management of the many potential complications. Patients with aortic aneurysm often have serious pathology of critical organs that remains uncorrected after the aortic operation. Bleeding, renal failure, myocardial infarction, and limb ischemia are the main concerns in the first 24 postoperative hours. In the first few days, pneumonia, urinary tract infection, catheter sepsis, colon ischemia, fluid overload, and electrolyte/acid-base imbalance are frequent additional concerns. After several days, wound complications, gastrointestinal hemorrhage, nutritional depletion, deconditioning, mental status changes, venous thromboembolism, and pressure ulcers come into play. While many of these complications become clinically manifest later, earlier attention can arrest the problems at a subclinical level. Long-term complications include ventral hernia, retained aneurysm growth, prosthetic graft infection, aortoenteric fistula, graft limb occlusions, and subsequent development of an aneurysm.

Follow-Up

Patients are seen frequently postoperatively to monitor the above complications. After recovery, they are seen annually for clinical exam and to reinforce lifestyle modification. A CT scan is done every 5 years to detect subsequent aneurysms.

95

Chapter 96

Endovascular Repair of Infrarenal Aortic Aneurysm

Jon S. Matsumura

Indications

The indications for endovascular repair of infrarenal aortic aneurysm are abdominal aortic aneurysm that is 5.5 cm or larger, enlarging 5 mm/6 months, symptomatic, or associated with other risk factors for rupture. Suitable anatomy for available endovascular devices.

Operative Principles

Endovascular repair requires extensive preoperative imaging and judgement to assess for anatomic qualifications, anticipate potential intraoperative problems, and select appropriate endovascular devices and sizes. Additional counseling and selection of patients is necessary to insure long-term follow-up and clinical success. Fluoroscopic guidance is used during the procedure and may be supplemented by intravascular ultrasound. There are several device-specific directions for use and potential pitfalls that are learned through extensive training and experience, which are not covered in this brief description.

Preop

The patient is given a mild, nondehydrating bowel laxative the day before. One gram of cefazolin is given intravenously. Two units of packed red blood cells are cross-matched. A cell saver collection is attached to the suction but is only processed if there is a need for transfusion. The abdomen and groins are shaved.

Procedure

Step 1. Supine position, bladder catheter, check fluoroscopy field is not obstructed, and prep from nipples to the knees.

Step 2. Femoral artery exposed through transverse incision above inguinal skin crease. The external iliac artery is double looped with two loops and the common femoral artery and branches controlled. Percutaneous access may be used in selected cases with some devices.

Step 3. A separate stab wound for sheath placement may be useful to minimize sheath kinking.

Step 4. Retrograde cannulation of the aorta is performed with an atraumatic guidewire on each side. A short angled guide catheter may be useful if the iliac arteries are tortuous or diseased and to exchange for a stiff guidewire. Hemostatic sheaths of appropriate size are inserted and pressurized flush bags attached. Iliac

---

Northwestern Handbook of Surgical Procedures, edited by Richard H. Bell, Jr. and Dixon B. Kaufman. ©2005 Landes Bioscience.

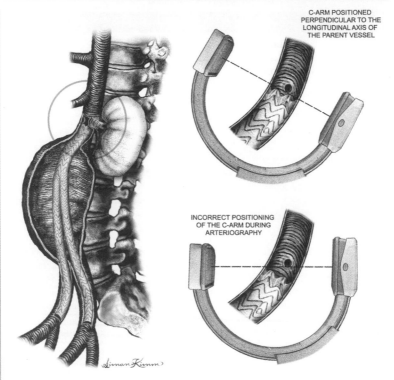

C-ARM POSITIONED
PERPENDICULAR TO THE
LONGITUDINAL AXIS OF
THE PARENT VESSEL

INCORRECT POSITIONING
OF THE C-ARM DURING
ARTERIOGRAPHY

96 Figure 96.1. Endovascular repair of infrarenal aoric aneurysm. Anatomy.

injury should be suspected if there is difficulty inserting the larger sheaths, and occasionally direct exposure is necessary for repair and access.

Step 5. Arteriography is performed to identify and assess branch arteries and measure intraluminal distances. A reassessment of the anatomic suitability for endovascular repair is performed. The procedure may be continued, aborted, or converted to open repair as necessary. Confirmatory arteriography—with the C-arm positioned perpendicular to the longitudinal axis of the parent vessel and the origin of the important branch vessel—may be performed just prior to each stage of deployment.

Step 6. Systemic heparin (5000 units) may be given if the patient has minimal bleeding.

Step 7. Unibody device deployment. The contralateral pull wire is inserted on the ipsilateral side and snared from the contralateral side. The device is inserted, checked for torque, and the jacket retracted. The proximal end is deployed and seated in place with an aortic balloon. The contralateral limb and the ipsilateral limb are similarly placed. Care is taken to select the balloon size, position and inflate properly. The deployment system is removed.

Step 8. Modular device deployment. The main trunk component is placed first with the contralateral leg hole positioned so that it can be easily cannulated. After this is verified, the contralateral sheath is advanced, and the contralateral leg is inserted and deployed. Extenders are used to establish long overlap and seal zones.

Step 9. After completion arteriography (and pressure measurements as indicated) shows suitable endograft deployment and configuration, the sheaths are removed. Often a retrograde sheath injection to assess the distal iliac anatomy is also performed.

Step 10. Verify hemostasis of all suture lines. Palpate for femoral pulses.

Step 11. Close incisions with absorbable suture in layers and skin with the knots buried.

Step 12. Check distal pulses/Doppler indices prior to moving patient off the operating room table.

Postop

Monitor organ function, particularly renal function. Maintain treatment of chronic medical comorbidities. Prescribe appropriate pain control. Check chest X-ray if central line attempted.

Complications

Bleeding, renal failure, myocardial infarction, and limb ischemia are the main concerns in the first 24 postoperative hours.

Follow-Up

Clinical outcomes after technically successful endovascular repair are dependent on long-term clinical and radiographic follow-up with selective reintervention. Long-term abnormalities and complications include endoleak (continued blood flow in the sac), aneurysm growth and rupture, device migration, device material failure, prosthetic graft infection, graft limb occlusion, and subsequent development of a new aneurysm.

96

Chapter 97

Aortofemoral Bypass for Obstructive Disease

Mark K. Eskandari

Indications

Indications include extensive bilateral iliac disease and the presence of buttock or thigh claudication.

Preop

The patient is assessed using a standard angiogram or a high-quality magnetic resonance angiogram to delineate the level of the infrarenal aorta and rule out a potential aneurysm and also to delineate the distal targets, typically the common femoral artery at its bifurcation. The patient is treated preoperatively with intravenous antibiotics.

Procedure

Step 1. The patient is placed in a supine position.

Step 2. The first two incisions are made obliquely in each of the groins, isolating the common femoral artery, superficial femoral artery, and profunda femoris artery bilaterally.

Step 3. Through a midline laparotomy incision, the infrarenal aorta is approached and carefully dissected at the level just below the renal arteries.

Step 4. The aorta-bifemoral bypass graft may be sewn in either an end-to-end or end-to-side fashion. End-to-end is used in cases where the extent of external iliac disease is minimal, allowing retrograde flow into the pelvis. An end-to-side anastomosis is utilized in cases where there is an extremely small aorta, typically in middle-aged women, or in cases where there is extensive external iliac disease, not allowing adequate retrograde perfusion to the pelvis. In both situations, the anastomosis is performed below the renal arteries.

Step 5. The tunnels are performed for the limbs of the graft from the groin up into the pelvis, along the native vessels underneath the ureter.

Step 6. The patient receives heparin systemically, typically 5000 units.

Step 7. The aorta is controlled proximally and distally. Typically complete aorta cross-clamp is necessary because a side clamp is very difficult to place due to the small size of the aorta. A longitudinal arteriotomy is fashioned and an end-to-side anastomosis with a bifurcated graft is performed with a 2-0 polypropylene suture. Alternatively, if an end-to-end anastomosis is performed, it is done in a standard fashion and the distal stump is oversewn with a running mattress and running over-and-over suture of 2-0 polypropylene.

Northwestern Handbook of Surgical Procedures, edited by Richard H. Bell, Jr. and Dixon B. Kaufman. ©2005 Landes Bioscience.

Step 8. The limbs of the graft are tunneled down to the level of the groins. After proximal and distal control of the femoral arteries is attained, the anastomosis is sewn to the bifurcation, not above it.

Step 9. The clamps are removed and flow is restored. Hemostasis is obtained.

The retroperitoneum is closed in layers, the abdominal cavity is closed, and the groins are closed in the standard fashion.

Postop

Postoperative considerations include monitoring of cardiac function, fluid requirements, and assessment of renal functions.

Complications

Complications include bleeding from the anastomosis or from the retroperitoneum requiring reoperation and pelvic ischemia from inadequate collateral flow resulting in colonic ischemia. Other potential long-term complications are an aortic-duodenal fistula as well as limb occlusion of one of the limbs of the prosthetic graft.

Follow-Up

Patients are usually discharged 1 week following the operation and should be seen back in the outpatient setting in 3 weeks to assure normal wound healing. The patient should be monitored with duplex ultrasound after 6 months and yearly thereafter.

97

Chapter 98

Axillofemoral Bypass

Joseph R. Schneider

Indications

Aortobifemoral bypass remains the preeminent operation for bilateral aortoiliac "inflow" disease, but axillofemoral bypass may be more appropriate for patients who are at high operative risk due to medical comorbidities, patients with "hostile abdomen," or acute lower extremity ischemia. Axillofemoral bypass with ligation of the infrarenal aorta for mycotic aneurysm, infected aortic graft, or aortoenteric fistula is the most frequently employed approach to these unusual clinical problems. Results in such patients may be much different from those in patients with primary operations for lower extremity ischemia, and the reader should bear this in mind when examining the original literature of axillofemoral bypass. In general, patients should be considered for axillofemoral bypass only if (1) they have severe bilateral aortoiliac occlusive disease, are considered inappropriately high risk for aortofemoral bypass, and are not candidates for percutaneous transluminal balloon angioplasty of at least one "inflow" side or (2) they suffer from one of the other unusual problems listed above.

Operative Principles

The basic concept of axillofemoral bypass is one healthy axillary artery can supply the ipsilateral arm and one or both legs, at least at rest. Grafts are placed subcutaneously from one infraclavicular area to the (nearly always ipsilateral) groin and in the case of a bifemoral reconstruction, from one groin across to the other, again subcutaneously. Placement of the axillary end of the graft anterior or posterior to the pectoralis minor muscle is probably unimportant with respect to results. It is far more important to place the axillary-graft anastomosis as medially as possible. Medial placement minimizes tension on the anastomosis with abduction of the arm and eliminates the need to divide the pectoralis muscle in most cases. The axillofemoral graft must be tunneled in the midaxillary line to prevent kinking of the graft with torso flexion or a kink over the costal margin which tends to be more prominent anteriorly than in the midaxillary line.

We usually place the axillofemoral graft first and then perform a side axillofemoral limb to the proximal end of femorofemoral limb anastomosis as low as possible. This so-called inverted "C" configuration of the femorofemoral limb maximizes flow in the majority of the axillofemoral limb. On some occasions, particularly when the femoral anastomosis is to the deep femoral artery, we perform the femorofemoral graft first and then perform an end axillofemoral limb-to-side femorofemoral limb anastomosis.

Slight redundancy of the femorofemoral limb and locating the femorofemoral graft tunnel well superior to the pubis reduce the risk of kinking at the heels of the

Northwestern Handbook of Surgical Procedures, edited by Richard H. Bell, Jr. and Dixon B. Kaufman. ©2005 Landes Bioscience.

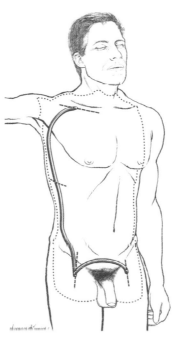

Figure 98.1. Axillofemoral bypass with femorofemoral anastomosis.

anastomoses. Unsupported Dacron, externally supported Dacron, expanded polytetrafluoroethylene (ePTFE), externally supported ePTFE, and even vein grafts have been used for axillofemoral bypass. There is no convincing evidence that any of these is superior to the others. However unproved, external support has "face validity" and ePTFE is probably more convenient since it is not associated with bleeding in the subcutaneous tunnel. I routinely use an 8 mm axillofemoral component and a 6 mm femorofemoral component, both externally supported ePTFE or, more recently the Gore-Tex® intering graft.

98

Preop

Optimization of cardiopulmonary function prior to axillofemoral bypass is desirable when possible. However, patients selected for this operation are generally suffering from advanced comorbidities and only limited "tuning" is possible in most. Preoperative control of infection including surgical treatment of foot sepsis when appropriate, is also important. We insist on a triphasic brachial or axillary artery Doppler waveform in the prospective donor limb, and the upper extremity with the highest blood pressure is most appropriate for graft origination. Preoperative baseline noninvasive lower extremity arterial studies are obtained, and adequate aortography and "runoff" angiography are necessary for operative planning. Aortic arch and subclavian arteriograms are obtained only if there is any question about the axillosubclavian arterial system or routinely if the patient's arteriogram is performed via the brachial or axillary artery.

Procedure

Step 1. The operation can be performed with predominantly regional anesthesia, but it is difficult to obtain adequate regional anesthesia to allow axillary dissection and graft tunneling. For this reason, general anesthesia is nearly always used in practice. Patients are placed supine, usually with a roll under the torso on the side of the axillofemoral limb to lift that side a little above the operating table. Most surgeons perform this operation with the arm on the side of the axillofemoral limb abducted to 90°. Any arterial monitoring catheters should be placed in the arm opposite the planned axillary artery incision. Use of two operating teams can expedite the operation when possible. The neck, chest, axilla, abdomen, and both groins and anterior thighs are prepped and draped.

Step 2. Transverse incision, generally about 8 cm long, is made inferior to the clavicle. The pectoralis major fibers are separated exposing the pectoralis minor muscle and the fat in which the axillary artery will be located medial to this muscle, inferolateral to the clavicle. Isolate the axillary artery from the pectoralis minor muscle laterally to the clavicle medially. The resident familiar with the common femoral artery must be cautioned that the axillosubclavian artery is much less resilient than the femoral artery, it is easily damaged during dissection, and injury to this artery, particularly medially where it may be difficult to obtain control and repair the artery due to the presence of the clavicle, may lead to substantial bleeding, frustration, and embarrassment.

Step 3. Femoral arteries are generally exposed and controlled through standard longitudinally oriented groin incisions.

Step 4. Create subcutaneous tunnels for the axillofemoral and femorofemoral grafts. Axillofemoral tunneling has been facilitated by the 65 cm long Gore tunneler that allows tunneling from the ipsilateral groin to the axillary incision without an intermediate (flank) incision. Care must be taken to position the axillofemoral tunnel in the midaxillary line, not to injure the neurovascular structures of the axilla when passing the tunneler through the axilla and not to pass the tunneler into the pleural space. Tunneling of the femorofemoral component may be performed with a tubular tunneler or a large clamp.

Step 5. The axillary artery is controlled, usually with a pediatric Cooley clamp medially and an angled (DeBakey) clamp laterally. Femoral vessels are controlled with clamps and vessel loops as appropriate.

Step 6. A longitudinal arteriotomy is made in the axillary artery. A longitudinal arteriotomy is generally started in each common femoral artery and may be carried across the deep or superficial femoral origins as appropriate.

Step 7. Graft limbs are spatulated and end-to-side anastomoses created with running suture, usually 5-0 polypropylene or CV-5 Gore-tex. One must remember to provide a little redundancy in these grafts to prevent "axillary pullout" and kinking of the femorofemoral limbs. An anastomosis must also be created between the axillofemoral and femorofemoral limbs. Geometry may be facilitated by placing the femorofemoral graft first and then placing an anastomosis between the distal end of the axillofemoral graft limb and the side of the femorofemoral graft just above the femoral anastomosis.

Step 8. The system is flushed by brief release of clamps and vessel loops prior to completion of anastomoses, the anastomoses are then completed, and clamps removed.

Step 9. Check for hemostasis and make certain using the Doppler that arterial flow is enhanced in the outflow and that perfusion is restored after removal of clamps in the "donor" arm.

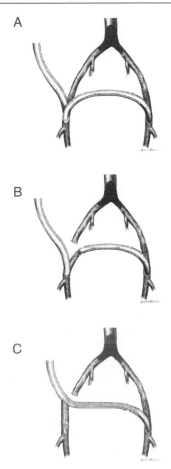

98

Figure 98.2. A) View of the inferior portion of a right axillobifemoral graft with the aorta, iliac, and femoral arteries behind it. This is the most common axillobifemoral bypass configuration with the axillofemoral limb to proximal (in this case right) femorofemoral limb anastomosis placed as distally as possible to maintain maximum flow through the entire axillofemoral limb. B) Division of the (in this case right) common femoral artery with end common femoral to end proximal femorofemoral limb anastomosis (Blaisdell). C) Division of the common femoral artery and side axillofemoral to end common femoral anastomosis on the axillofemoral limb side and continuation of the same graft to the contralateral femoral artery (Rutherford). Configurations described in B and C theoretically both maximize flow through the axillofemoral portion of the graft and prevent competitive inflow from the iliofemoral system on the axillofemoral side. The latter configuration requires only three anastomoses instead of the four required by configurations in A and B.

Step 10. Close wound in layers.

Postop

Patients may eat and resume activity as soon as they recover from anesthesia. This operation is a relatively mild injury to most patients. While they might otherwise be ready for discharge on the second or even first postoperative day, hospital stay is generally determined by comorbidities more than the trauma of surgery. Axillofemoral bypass is an inferior operation with respect to hemodynamic performance. Axillofemoral bypass results in an ankle-brachial-index of <0.7 with perfect outflow, whereas an aortofemoral bypass would produce an ankle-brachial-index of 1.0. It is not clear that this inferior performance is related to occult axillosubclavian inflow disease, inability of a *normal* axillosubclavian system to provide adequate flow to an arm plus one or both legs, or the effects of the increased resistance of smaller diameter longer grafts, but there is likely some contribution from each of these causes. This operation is virtually never appropriate for claudication, with the possible exception being the patient with the so-called "hostile abdomen." Such patients can often be treated with balloon angioplasty or some other procedure. However, this operation provides adequate enhancement of perfusion to provide limb salvage in most patients.

Complications

Axillofemoral bypass is a much less significant injury to the patient than an aortofemoral bypass would be. However, patients are generally selected for this operation because they are unacceptable risks for aortofemoral bypass. Thus, mortality reported in series of axillofemoral bypass may be substantially higher than that observed after aortofemoral bypass. Death is most often due to myocardial infarction or other cardiac morbidity. Graft infection is uncommon, but problematic when it occurs. Axillary pullout syndrome has been described as a disruption of the axillary artery to graft anastomosis associated with abduction of the arm, implying that the graft was probably made too short, and is the inspiration for the recommendation to make this graft a little redundant. Thrombosed grafts have often been the apparent source of emboli to the leg or, more often, the arm on the side of the axillary anastomosis. Axillofemoral bypass is more likely than aortofemoral bypass to fail (thrombose) during the postoperative period. Primary patency of axillofemoral grafts should be about 60-70% at 5 years, whereas primary patency of aortofemoral grafts should be 85-95% at 5 years. Graft thrombosis is the most common long-term complication, but axillofemoral prosthetic grafts are relatively easily thrombectomized, generally with local anesthesia. Patients are advised not to sleep on the side of the axillofemoral graft limb to try to reduce the chance of graft compression, stasis, and ultimate thrombosis, although it has been difficult to demonstrate that lying on the side of the axillofemoral graft limb produces graft compression.

Follow-Up

We recommend postoperative lower-extremity noninvasive arterial studies to determine whether distal perfusion has been significantly improved compared to preoperative studies. We recommend a postoperative office visit 2-3 weeks after discharge from the hospital. There is no convincing evidence that noninvasive graft surveillance can identify impending failure of axillofemoral grafts. Therefore, once wounds are healed, we recommend no more than yearly scheduled follow-up visits.

Chapter 99

Femorofemoral Bypass

Joseph R. Schneider

Indications

Femorofemoral bypass is a common approach to predominantly unilateral iliac artery occlusive disease. Advances in percutaneous endoluminal therapies have made it possible to treat many patients without open surgery. On the other hand, improvements in medical care and the relentless aging of the population in developed countries has meant that many patients with advanced medical comorbidities who would previously have died now survive long enough to present with advanced symptoms of lower extremity arterial occlusive disease, some of them with anatomy appropriate for femorofemoral bypass.

This operation is rarely appropriate for claudicants because of inferior hemodynamic performance and patency compared to aortofemoral bypass. However, this is an important and adequately effective less invasive therapy for many high-risk patients with predominantly unilateral iliac artery occlusive disease, as one method of salvage after thrombosis of one limb of an aortobifemoral graft, and occasionally in patients with mycotic aneurysms or other nonocclusive iliac artery pathology. Finally, this operation may be required to complete the reconstruction after aortouniiliac endoluminal graft placement for aortic aneurysm.

Operative Principles

The basic concept of femorofemoral bypass is that one healthy iliac artery can supply both legs. Grafts are placed in a suprapubic subcutaneous position from one "donor" femoral artery across to the other. The grafts are usually placed in a so-called inverted "C" configuration. Slight redundancy of the femorofemoral limb and location of the graft tunnel well superior to the pubis tend to prevent compression and to reduce kinking at the heel of the femoral anastomoses. Just as with aortofemoral bypass, it is critical that good outflow is assured, and we embrace the classic view that the deep femoral artery is the most important target artery in nearly all "inflow" procedures.

Dacron grafts were probably the first prosthetic grafts used for femorofemoral bypass. Externally supported Dacron, expanded polytetrafluoroethylene (ePTFE), and externally supported ePTFE grafts have subsequently been used with each being touted as the solution to the problems of femorofemoral bypass just as they have with axillofemoral bypass. There is no convincing evidence that this is the case. I nearly always choose an externally supported 6 mm diameter ePTFE graft or, more recently, a Gore-Tex® intering graft for femorofemoral bypass. Greater saphenous or other vein may be an appropriate conduit when infection is present or an unusually high risk, e.g., in patients with intestinal stomas or remote infection.

Northwestern Handbook of Surgical Procedures, edited by Richard H. Bell, Jr. and Dixon B. Kaufman. ©2005 Landes Bioscience.

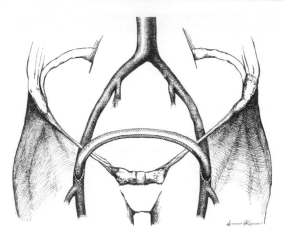

Figure 99.1. Femorofemoral bypass.

Preop

Patients selected for femorofemoral bypass are frequently, but not always, burdened with advanced medical comorbidities, and some optimization of cardiac and pulmonary function may be possible in some of these patients. Preoperative control of infection including surgical treatment of foot sepsis when appropriate is also important. Preoperative baseline noninvasive lower extremity arterial studies are obtained and adequate aortography and "runoff" angiography are necessary for operative planning. Particular attention must be paid to both the noninvasive and angiographic information regarding the potential donor limb. Patients with anything but normal common and external iliac arteries on angiogram may need "pull-back pressures" across any areas of concern. This may require augmentation with an arterial dilator such as papaverine. The consequence of depending on a less than optimal donor iliac arterial system may be symptoms of steal in the donor side leg, inadequate enhancement of perfusion of the recipient leg, early graft thrombosis, and any combination of the above. If a hemodynamically significant lesion is identified, it must be treated (usually by balloon angioplasty), and if it cannot be treated an alternative method of reconstruction should be chosen.

Procedure

Step 1. The lower abdomen and both groins and anterior thighs are prepped and draped. Use of two operating teams can expedite the operation when possible. General, spinal, or even local anesthesia may occasionally be used.

Step 2. The femoral arteries are exposed and controlled through standard longitudinally oriented groin incisions.

Step 3. Subcutaneous tunneling of the femorofemoral graft (performed with a tubular tunneler or a large clamp) in a suprapubic position. Remember to place the tunnel well superior to the femoral arteries and to provide a little redundancy in the grafts to prevent kinking at the heel of each anastomosis.

Step 4. Heparin administered, femoral vessels controlled with clamps and vessel loops as appropriate.

Step 5. A longitudinal arteriotomy is generally started in each common femoral artery and may be carried across the deep or superficial femoral origins as appropriate.

Step 6. Graft limbs are spatulated, and end-to-side anastomoses created with running suture, usually 5-0 polypropylene or CV-5 Gore-tex.

Step 7. Flush by brief release of clamps and vessel loops prior to completion of anastomoses, complete anastomoses, remove clamps.

Step 8. Check for hemostasis, make certain using the Doppler that arterial flow is enhanced in the recipient outflow and remains qualitatively unchanged in the donor leg outflow arteries.

Step 9. Close the wounds in layers.

Postop

Diet and activity can resume as soon as the patient recovers from anesthesia. This operation is a relatively mild injury to most patients, and while they might otherwise be ready for discharge on the second or even first postoperative day, hospital stay is generally dictated by comorbidities.

Femorofemoral bypass is an inferior operation with respect to hemodynamic performance and results in a predicted ankle-brachial-index of 0.82 with perfect outflow, whereas aortofemoral bypass would produce an ankle-brachial-index greater than 1.0. It is not clear that this inferior performance is related to inability of a *normal* iliac system to provide adequate flow to both legs, inadequate appreciation of the effects of donor iliac atherosclerotic occlusive disease, or the effects of the increased resistance of smaller diameter longer grafts, but our examination suggested some contribution from each of these causes.

Complications

Femorofemoral bypass is a much less significant injury to the patient than an aortofemoral bypass would be. However, patients are often, although not always, selected for this operation because they are at unacceptable risk for aortofemoral bypass. Thus, mortality reported in series of femorofemoral bypass may be substantially higher than that observed after aortofemoral bypass. Death is most often due to myocardial infarction or other cardiac morbidity. Graft infection is uncommon, but problematic when it occurs. Femorofemoral bypass is more likely than aortofemoral bypass to fail (thrombose) during the postoperative period. Primary patency of femorofemoral grafts should be about 60-70% at 5 years, whereas primary patency of aortofemoral grafts should be 85-95% at 5 years. Graft thrombosis is the most common long-term complication, but thrombosed femorofemoral prosthetic grafts can often be successfully thrombectomized under local anesthesia.

Follow-Up

We recommend postoperative lower-extremity noninvasive arterial studies to determine whether distal perfusion has been significantly improved compared to preoperative studies. We recommend a postoperative office visit 2-3 weeks after discharge from the hospital. There is no convincing evidence that noninvasive graft surveillance can identify impending failure of femorofemoral grafts. Therefore, once wounds are healed, we recommend no more than yearly scheduled follow-up visits.

Chapter 100

Femoral-Popliteal Bypass with a Vein or Prosthetic Graft

John V. White and Christopher Bulger

Indications

The most common indication for femoral-popliteal bypass is claudication that has not responded to exercise therapy. This procedure may also be of benefit for patients with limb-threatening ischemia who have a patent popliteal artery and good collateral run-off.

Preop

All patients who are to undergo a femoral-popliteal bypass should have a complete imaging study of the arterial tree from the proximal abdominal aorta to the toes. This may be a conventional angiogram, a magnetic resonance angiogram, or a CT angiogram. Duplex imaging of the greater saphenous vein is also helpful to determine whether this is adequate for the proposed surgery. Since a reduced ankle-brachial index is associated with coronary artery disease, patients should also undergo a cardiac evaluation prior to surgery.

Procedure

Step 1. The procedure may be performed under general, epidural, or local anesthesia. The patient is placed in the supine position with the leg flexed slightly at the knee and externally rotated. A curvilinear incision is made over the femoral pulse superiorly extending medially to permit exposure of the saphenous vein.

Step 2. The proximal groin incision is deepened to expose the distal 3 cm of the common femoral artery. The vessel is isolated from adjacent tissues and surrounded with vessel loops proximally and distally.

Step 3. The popliteal artery is exposed through a medial incision just proximal to the level of the femoral condyle. The incision is deepened carefully to prevent injury to the greater saphenous vein. The fascia is incised to open the popliteal space. The neurovascular bundle is identified, the artery is separated from the vein and nerve, and a 3 cm segment of vessel beyond the occlusive disease is isolated and surrounded by vessel loops proximally and distally.

Step 4. The saphenous vein is identified at the saphenofemoral bulb, which is generally at the level of the bifurcation of the common femoral artery. The saphenofemoral junction is clamped proximally and distally and divided. The stump leading to the common femoral vein is oversewn with a 6-0 polypropylene suture. The tributaries of the greater saphenous vein are ligated and divided. For the construction of an in situ bypass, the required length of the vein is exposed through a continuous incision. The tributaries are simply ligated along the length of the vein,

Northwestern Handbook of Surgical Procedures, edited by Richard H. Bell, Jr. and Dixon B. Kaufman. ©2005 Landes Bioscience.

and the valves are cut using a valvulotome to render them functionless. If the vein is of adequate diameter at the distal end, it may be removed through small, discontinuous incisions after its tributaries are ligated and divided. The vein is reversed, so that the valves do not interfere with blood flow, and placed through a tunnel following the course of the superficial femoral artery as described in Step 5. If a prosthetic graft is to be used, exposure of the greater saphenous vein is not undertaken.

Step 5. If either a reversed saphenous vein or a prosthetic graft is used, a subsartorial tunnel adjacent to the course of the superficial femoral artery is created through which the graft is passed. This is usually performed with a tunneling instrument which has a removable tip and a hollow core through which the reversed saphenous vein or prosthetic graft is passed without twisting or kinking.

Step 6. The patient is given 5000 U of heparin intravenously to prevent blood clotting in the arteries during the period of occlusion for the construction of the anastomoses.

Step 7. The distal common femoral artery is occluded proximally and distally to isolate a 3 cm segment. The anterior wall is opened with an 11 blade and a small button of the anterior wall is removed with an angled Potts arteriotomy scissors. The proximal end of the vein or prosthetic graft is beveled and sewn end-to-side to the common femoral artery with a 5-0 polypropylene suture. Once the anastomosis is completed, a vascular clamp is placed across the origin of the graft and the clamps on the common femoral artery are removed.

Step 8. The popliteal artery is then clamped proximally and distally to isolate a 3 cm segment. The superior surface is opened with a scalpel and a small button of the arterial wall is removed with a Potts scissors. The distal end of the graft is shortened and beveled to the appropriate length and shape and sewn end-to-side to the artery with a 5-0 or 6-0 polypropylene suture. Just prior to the completion of the anastomosis, the graft and the artery are flushed of air and debris. The anastomosis is completed and secured. The clamps are removed from the origin of the graft and the proximal artery. After several seconds, the distal clamp is removed to initiate flow to the distal leg.

Step 9. The technical precision of the distal anastomosis and outflow to the lower leg is confirmed by angiography performed by injecting contrast directly into the graft after insertion of a 20-gauge angiocatheter into the proximal portion and occluding the inflow into the graft. If an in situ graft has been constructed, the X-ray is used to confirm that all of the valves have been completely lysed.

Step 10. After the radiograph is reviewed and technical precision is confirmed, hemostasis is secured, the wounds are irrigated, and the subcutaneous tissues are reapproximated with a synthetic absorbable suture. The skin is closed with either staples or a synthetic absorbable suture.

Complications

As with any surgical procedure, there can be both systemic and local complications associated with femoral popliteal bypass. The major systemic complication is myocardial infarction. This occurs in approximately 5% of patients undergoing such procedures. Additionally, transient renal insufficiency may be associated with the administration of contrast for the preoperative or completion angiograms performed to evaluate arterial anatomy and bypass. Local complications are wound-related problems, graft thrombosis, and infection. Wound-related problems are infection and the formation of seromas. Seromas are most common in the groin and may lead to infection of the groin. Any wound infection raises the risk of graft infection. Graft infection may be a significant problem, especially with prosthetic grafts. When involving the anastomosis, the graft infection may require graft removal and repair of the underlying artery. Graft thrombosis, which is more common with prosthetic graft than with autogenous vein graft material, requires graft thrombectomy. This can be performed by reopening the incision and gaining control of the graft a short distance away from the anastomoses. A thrombectomy catheter then may be inserted and the clot removed. Evaluation of the graft and the anastomoses with either angiogram or angioscope must be performed. The underlying cause of the thrombosis must be sought and corrected. Bleeding from the graft or anastomosis is also a complication. With prosthetic grafts this is most likely to occur in the area of the anastomoses. In vein grafts, this may occur either from the anastomoses or from a side branch which had been ligated. The source of the bleeding must be identified and surgically corrected.

Postop

Patients spend overnight in an intermediate care unit of the hospital capable of monitoring bleeding, clotting of the graft, or infection. They are allowed to be out of bed if they are comfortable. An appropriate regular diet may be resumed after surgery. Antibiotics are continued for one or two doses after surgery for wound prophylaxis or longer if there is a distal ulcer or infection. Antiplatelet or anticoagulation therapy may be initiated based upon the surgeon's perception of the risk of graft thrombosis.

Follow-Up

Patients are seen in the surgeon's office at 1 week, 1 month, 3 months, and then every 6 months for an evaluation of their walking ability and graft patency.

Chapter 101

Composite Sequential Bypass

William H. Pearce

Indications

A composite sequential bypass is indicated in patients with limb-threatening ischemia in whom there is not a sufficient length of vein to span the femoral tibial segment. Limb-threatening ischemia includes patients with nonhealing ulcers, ischemic rest pain, and gangrene. For infrainguinal bypass, an all-autogenous reconstruction (the greater saphenous vein or splicing veins from upper and lower extremities) functions better than prosthetic grafts. When these options are not available, the composite sequential offers a greater long-term patency than an all-prosthetic bypass.

Operative Principles

This operation is a combination of a prosthetic femoral-popliteal bypass and a distal venous graft to a tibial vessel. This hybrid graft consists of both a prosthetic and a venous segment. The venous segment is most commonly derived from the lesser saphenous vein of the ipsilateral leg. The operative exposures are those that are described previously for femoral popliteal bypasses and tibial bypasses. The operative procedure begins with the harvesting of the lesser saphenous vein. This is done with the patient supine and the leg elevated. This is the most difficult part of the procedure. Once the lesser saphenous vein is harvested, the femoral, popliteal, and distal tibial arteries are exposed. The intermediate anastomosis, that is the anastomosis between the vein and the prosthetic, is unique to this procedure.

This anastomosis is created with an end vein to side of prosthetic graft. This large anastomosis is perhaps the reason for success of this procedure. The large diameter of this anastomosis allows for intimal hyperplasia to occur without impeding the flow of the graft.

Preop

As with all patients with lower extremity arterial disease, a preoperative blood flow study and an arteriogram are required. Additionally, these patients also require complete vein mapping for all usable vein. Usable vein is marked preoperatively.

Procedure

Step 1. The patient is placed in the supine position on the operating table and, after general anesthesia, the leg is prepped and draped in a standard sterile fashion.

Step 2. The leg is elevated. The operation begins with harvesting the lesser saphenous vein at the previously marked site; the posterior skin overlying the lesser saphenous is opened. The vein is harvested from ankle to popliteal fossa. The wound is left open. If the vein is less than 3-4 mm in diameter it is not usable. Alternative sources of vein are then necessary (upper extremity).

Step 3. Exposure of the femoral artery, popliteal artery either above or below the knee, and the distal tibial vessel is accomplished using the techniques previously described.

Step 4. Next, the femoral popliteal bypass using prosthetic graft is completed initially. This procedure uses an 8 mm polytetrafluoroethylene (PTFE) graft following systemic heparinization with 5000 U of heparin. The anastomosis is created end-to-side in both locations.

Step 5. The venous segment can be implanted either in an orthograde or retrograde direction. The orthograde technique requires a proximal anastomosis to be completed first, with stripping of the valves using a Mills valvulotome. Using the reverse technique, this step is omitted and a proximal anastomosis is completed.

Step 6. The distal anastomosis is created to the tibial vessel again, using similar techniques as described above. Upon completion, the entire system is flushed of debris.

Step 7. An intraoperative completion arteriogram is performed to demonstrate technical adequacy of all anastomoses.

Step 8. Following completion of the arteriogram, hemostasis is achieved and the wounds are closed in multiple layers. It is important not to forget to close the posterior lesser saphenous vein harvest site.

Step 9. Occasionally, these patients require postoperative heparinization and as a result a number 10 Jackson Pratt drain is left in calf and mid-thigh incisions.

Postop

These patients are treated like those with any distal bypass. They are observed in the intensive care unit overnight for bleeding or early graft occlusion. Postoperative Doppler tones or ankle-brachial indices are performed every 2-4 hours. In general, most patients are discharged on aspirin. However, based on surgeon's preference, the patients may be discharged on Coumadin or other anticoagulants or antiplatelet agents.

Complications

Early graft failure may occur in approximately 1-10% of patients. In addition, postoperative hematomas are common, particularly in patients with perioperative anticoagulation. Postoperative graft infections are uncommon. Skin and wound necrosis is, however, a common problem occurring in 2-20% of patients. The posterior incision for the lesser saphenous vein is rarely a significant problem. Similar to all lower extremity revascularization, the preoperative mortality of between 2-6% occurs.

Follow-Up

Graft surveillance should be performed at 3-month intervals for the first year and at 6-month intervals thereafter.

Chapter 102

Infrapopliteal Bypass: Vein or Prosthetic

Thomas W. Kornmesser

Indications

The primary indication for a distal (tibial vessel) bypass is chronic critical ischemia of the extremity. This would include patients with rest pain or patients with nonhealing ulcers or gangrene. It may occasionally be necessary in acute thrombosis, e.g., acutely occluded popliteal aneurysm or an acute thrombosis of the femoral and/or popliteal vessels. Limiting claudication is generally not an indication for infrapopliteal bypass.

Preop

1. The patient must be fully evaluated preoperatively to be certain that all the co morbidities are controlled, i.e., diabetes, coronary artery disease, pulmonary disease, renal disease, and/or hypertension.
2. Accurate preoperative angiography must be obtained including both inflow (aorta and iliac) and outflow (including visualization of the foot vessels). Occasionally this is not possible and intraoperative angiography is necessary to delineate the distal circulation. It is also noted that when chronic elevation of the creatinine is present, it is well advised to use minimal amounts of contrast with good hydration. In addition, in this situation, the use of gadolinium as the contrast choice should be considered.
3. Assessment is then made to determine the outflow vessel to the foot.
4. The inflow vessel is usually the common femoral artery but can be the superficial femoral, deep femoral, or even the popliteal artery depending upon their quality and the length and quality of the conduit available.
5. Selection is then made of the conduit, which may well be one of the most important decisions. If necessary, duplex scanning of the vein can be helpful. The first choice is generally the ipsilateral greater saphenous vein. Secondarily, the contralateral greater saphenous vein can be used. Alternates also include the lesser saphenous vein, cephalic vein, basilic vein, or the brachial vein. The superficial femoral vein can also be used. The last choice would be a prosthetic graft (polytetrafluoroethylene [PTFE], Dacron). The cryogenic vein and the umbilical vein are also alternatives.
6. The selection of anesthesia is important and the most ideal would be epidural with the catheter being left in place 48-72 hours postoperatively. General anesthesia is necessary when an upper extremity vein is to be harvested.
7. Depending on the patient, an arterial line is helpful for monitoring. A central line is also helpful. Occasionally in patients with severe comorbidities a Swan-Ganz catheter is helpful intraoperatively and postoperatively.

Northwestern Handbook of Surgical Procedures, edited by Richard H. Bell, Jr. and Dixon B. Kaufman. ©2005 Landes Bioscience.

Procedure

Step 1. A vertical incision over the femoral triangle is made to obtain control of the common femoral, deep femoral, and superficial femoral arteries.

Step 2. If the greater saphenous vein is to be the conduit and it is decided to perform an in situ bypass, the saphenofemoral junction is isolated. A second team, if available, works distally to expose the greater saphenous vein. The entire vein is unroofed through a continuous incision.

Step 3. The selected outflow vessel is then dissected. The posterior tibial and the peroneal arteries are approached medially, taking down the gastrocnemius and soleus muscles. The anterior tibial artery is approached through the anterior compartment. When performing an in situ graft to the distal one third of the anterior tibial artery or the dorsalis pedis artery, the vein is taken over the tibia. When the target is in the upper two thirds of the anterior tibial artery, the vein is brought through the interosseous membrane. Generally, the approach to the anterior tibial artery with a reversed vein is through the interosseous membrane.

Step 4.

- When a reversed vein is used, a subsartorial tunnel is made between the groin wound and the calf wound with a tunneler. Often exposure of the distal popliteal fossa is necessary. The patient is heparinized.
- When an in situ graft is used, the patient is first heparinized, and the vein is taken down from the saphenofemoral junction with a small oval. The femoral vein is then repaired. It is important to maintain a long length of the large branches of the vein in the groin for angioscopy and arteriography. The first valve of the vein is excised under direct vision.

Step 5.

- When using a reversed vein, the distal anastomosis is often done first after threading the vein through the tunnel. A vertical arteriotomy is made after control has been obtained. The anastomosis is performed with a small running or interrupted stitch. The proximal anastomosis is then begun. Prior to completing the anastomosis, angioscopy and arteriography are done. The anastomosis is then finished.
- The in situ graft is begun in the groin with the anastomosis to the femoral artery. Flow is then established to the vein. The valves are cut with a valvulotome either blindly or under direct vision with an angioscope. The distal anastomosis is performed in the same manner as in a reversed vein bypass. Angioscopy and arteriography are performed through retained branches of the saphenous vein.
- The technique for use of a prosthetic graft is similar to a reversed vein. Suturing into a venous patch to help with the so-called compliance mismatch is often done. (Taylor patch)

Postop

The patient should have graft surveillance with a Duplex scan prior to discharge or shortly after and on a regular basis. Aggressive treatment of a defect can help prevent early and late graft occlusions.

Complications

Systemic complications can occur which are related to the patient's comorbidities, such as myocardial infarction or other cardiac dysfunction. Pulmonary problems

102

and/or renal dysfunction can occur. Early graft occlusion and/or distal embolization can be similarly problematic. Wound problems including infection are more common in the in situ graft because of the length of the incision and the dissection.

Follow-Up

It is imperative that there be close follow-up in the early postoperative period. The greatest percentages of failures occur within the first year. The time schedule does depend on the patient's hospital course and problems, but they should be seen at least within the first month of discharge if not sooner. If a surveillance duplex scan has not been done it should be done at this visit. This should be repeated at 3 months postoperatively and then at 6 month intervals postoperatively. The patients should be followed at least yearly after the first postoperative year and sooner if there appears to be a problem. Besides the duplex scan the evaluation should include but should not be limited to palpation of pulses, ankle pressures, and presence or absence of bruits over the reconstruction. The observations all need to be recorded. If abnormalities are found, one needs to consider arteriography and/or intervention.

102

Lower Extremity Thrombectomy/Embolectomy

Mark K. Eskandari

Indications

Indications for a lower extremity embolectomy or thrombectomy are acute occlusions of the lower extremity necessitating immediate revascularization.

Preop

Most patients who require this procedure have an acute occlusive event; therefore, they require systemic heparinization prior to their procedure and adequate resuscitation. Additional preoperative workup includes an assessment of their basic serum chemistry profile, including CBC, platelets, coagulation, and a chemistry panel. The patient should also be consented for a thrombectomy/embolectomy and possible revascularization. These procedures are typically done under general anesthesia. However, they also may be done under spinal or epidural anesthesia or local anesthesia.

Procedure

Step 1. The patient is on intravenous heparin at a continuous infusion and given a dose of intravenous antibiotics.

Step 2. With the patient in a prone position, the lower extremity is prepped from the umbilicus down to the toes bilaterally.

Step 3. Incision is made in the groin over the common femoral artery of the ischemic limb. The common femoral artery as well as the superficial femoral artery and profunda femoris artery are carefully dissected and encircled with vessel loops, taking care not to injure the crossing lateral circumflex vein, which is found at the crux of the femoral artery bifurcation.

Step 4. With the patient on heparin, proximal and distal control is obtained by clamping the common femoral artery, the superficial femoral, and the profunda femoris artery.

Step 5. A transverse arteriotomy is made in the common femoral artery near the femoral bifurcation.

Step 6. A Fogarty embolectomy catheter is passed first distally. Usually for the lower extremity, a 4 Fogarty embolectomy catheter is advanced with markers of 10 cm marked on the catheter. The catheter is passed down the superficial femoral artery as far as it can go and the balloon is carefully inflated with gentle back-tension retrieving the clot.

Step 7. A similar maneuver is performed after cannulating the profunda femoris artery. These steps are repeated until there is no retrieval of clot after two passages through both arteries.

Northwestern Handbook of Surgical Procedures, edited by Richard H. Bell, Jr. and Dixon B. Kaufman. ©2005 Landes Bioscience.

Step 8. The thromboembolectomy is performed through the proximal portion of the common femoral artery using the same techniques.

Step 9. At this point, the transverse arteriotomy is closed with interrupted 6-0 polypropylene sutures. Prior to the completion of the anastomosis, it is flushed proximally and distally, the anastomosis completed, and then the clamps removed sequentially.

Step 10. The clamps are first removed from the common femoral artery, the profunda femoris artery, and then the superficial femoral artery, restoring flow to the lower extremity.

Step 11. A completion angiogram is performed to delineate the adequacy of the thromboembolectomy.

Step 12. If this is inadequate or if there is residual clot in the tibial vessels, the next approach is to cut down on the below-knee popliteal artery through a medial incision.

Step 13. The incision is made just below the knee in the soft part of the medial aspect of the leg approximately 2 cm posterior to the posterior edge of the tibia. In making the incision, it is important to be careful of the greater saphenous vein traveling along this course.

Step 14. The incision is carried down to the level of the superficial fascia, which is incised. The gastrocnemius muscle is retracted posteriorly and the vascular fossa is identified.

Step 15. With careful dissection in this avascular plane, the below-knee popliteal artery is carefully identified as well as its paired popliteal veins. The artery is carefully dissected from the popliteal veins and encircled with vessel loops.

Step 16. A thromboembolectomy is performed using the Fogarty catheter as described above in a similar manner through transverse arteriotomy and repaired in a similar manner.

Step 17. A completion angiogram is performed, confirming the adequacy of the thromboembolectomy.

Step 18. Any residual clot of significance should be retrieved through repeat of the above-outlined steps.

Postop

The patient is extubated, taken to the intensive care unit, resuscitated, and maintained on anticoagulation. The source of thromboembolic disease is identified with additional testing, including ruling out cardioembolic source based on an EKG to rule out atrial fibrillation, a transesophageal echocardiogram to rule out mural thrombus within the heart, CT scan of the entire aorta and iliac system to rule out an arterial source for embolic disease and plaque rupture, and a hypercoagulable screen as well. The patient is maintained on anticoagulation until the determination of the acute ischemic event is delineated.

Complications

Complications associated with this procedure include trauma to the arteries resulting in a dissection, therefore necessitating a local bypass operation. Additional complications include rupture of either the balloon or the artery and arterial ruptures identified with the completion angiogram necessitating repair with an interposition vein graft; prolonged ischemic event prior to the thromboembolectomy, and the resultant development of acute compartment syndrome that may require fasciotomy at a later time.

103

Follow-Up

Patients are usually discharged on the day following the operation and should be seen back in the outpatient setting in 3 weeks to assure wound healing. Patients should be monitored with duplex ultrasound after 6 months and yearly thereafter.

Chapter 104

Repair Popliteal Aneurysm: Emergent (Thrombosed)

Mark K. Eskandari

Indications

Indications for repair of a popliteal aneurysm include thrombosis of the aneurysm or distal embolization resulting in thrombosis of the tibial vessels.

Preop

Diagnosis of a popliteal aneurysm may be made by duplex ultrasonography, CT scan, or MRI. An angiogram or high-quality magnetic resonance angiogram is typically required to delineate the proximal extent of the aneurysm and distal targets for a potential bypass. If no distal target is appreciated on an angiogram, thrombolysis may be utilized for 24 to 48 hours to open an acutely occluded tibial vessel.

Procedure

Step 1. Anesthetic options include spinal or general anesthesia.

Step 2. Intravenous antibiotics are administered.

Step 3. Two incisions are made over the popliteal artery. One is above the knee medially and one is below the knee medially. Through these incisions, the greater saphenous vein is harvested.

Step 4. The popliteal artery above the knee and below the knee is carefully dissected and circled with vessel loops.

Step 5. After obtaining control, the patient receives systemic heparinization, typically 5000 U IV.

Step 6. Proximal control is obtained in the above-knee popliteal artery.

Step 7. The bypass is performed in an end-to-side fashion between the greater saphenous vein, which is tunneled deep between the two heads of the gastrocnemius muscle, and the above-knee popliteal.

Step 8. Distal anastomosis is performed in a similar manner to an outflow vessel, typically the below-knee popliteal artery or one of the tibial arteries.

Step 9. The popliteal artery below the proximal anastomosis is ligated using an umbilical tape or is oversewn with a polypropylene suture.

Step 10. The distal below-knee popliteal artery proximal to the distal anastomosis also is ligated using an umbilical tape or is oversewn with a polypropylene suture. This excludes direct perfusion to the aneurysm.

Step 11. A completion angiogram is performed.

Alternatives to this procedure include a posterior approach with the patient in a prone position. This is typically not necessary in cases of an emergent thrombosis of the popliteal artery.

Northwestern Handbook of Surgical Procedures, edited by Richard H. Bell, Jr. and Dixon B. Kaufman. ©2005 Landes Bioscience.

Postop

Postoperative considerations include the adequacy of revascularization. No anticoagulation is necessary with a vein bypass graft.

Complications

Complications include an inability to do a distal bypass necessitating an amputation; wound infections; absence of a vein for bypass necessitating a prosthetic graft; or bypass ultimately requiring anticoagulation for the prosthetic bypass graft.

Follow-Up

Patients are usually discharged 4-7 days following the operation and should be seen back in the outpatient setting in 3 weeks to assure normal wound healing. The patient should be monitored with duplex ultrasound after 6 months and yearly thereafter.

Chapter 105

Exploration for Postoperative Thrombosis

Thomas W. Kornmesser

Indications

Exploration for postoperative thrombosis is indicated for sudden acute thrombosis of recent bypass surgery and early thrombosis usually secondary to a technical problem.

Preop

Close postoperative observation is necessary to pick up early problems. It is advisable to begin anticoagulation with heparin as soon as the diagnosis is made to protect as much runoff as possible. An attempt to find any sign of a coagulopathy with appropriate laboratory data is necessary. It is necessary to diagnose and correct any medical comorbidity that could make surgical intervention prohibitive. Catheter-directed thrombolysis is generally not considered because of the proximity of the past surgery. Occasionally preoperative angiography is helpful, but the time necessary to perform it may jeopardize the extremity. In addition, the amount of contrast and the type of contrast may put the kidneys at risk in these very sick patients.

Procedure

Step 1. Exploration of the bypass either proximally or distally depending on where the problem is thought to be.

Step 2. Thrombectomy is performed with a Fogarty catheter. Occasionally intraoperative thrombolysis can be used with direct administration into the distal circulation as an adjunct.

Step 3. Intraoperative arteriography (be aware of kidney function) and/or angioscopy are done to delineate the cause of the thrombosis.

Step 4. Repair of the cause of the thrombosis is attempted, including patch angioplasty, lyses of a retained valve, ligation of arteriovenous fistula, or perhaps a repair of an anastomosis. If this cannot be done and the problem is thought to be the quality of the bypass graft, consider replacement of the graft preferably using autogenous tissue. A prosthetic graft can be used if nothing else is available or the medical condition dictates.

Step 5. Completion arteriography (be aware of kidney function), angioscopy, and/or duplex scanning are done to be certain the problem has been repaired.

Step 6. Depending upon the duration and severity of ischemia, a fasciotomy should be considered. This can be a formal four-compartment fasciotomy or a limited procedure such as a subcutaneous fasciotomy limited only to the anterior compartment. The decision is made clinically by placing an angiocath into the

Northwestern Handbook of Surgical Procedures, edited by Richard H. Bell, Jr. and Dixon B. Kaufman. ©2005 Landes Bioscience.

compartment and attaching it to a transducer which easily permits compartment pressure measurements.

Postop

Maintain anticoagulation with heparin and/or antiplatelet drugs. Close control is necessary to help prevent wound problems. Early graft surveillance is mandatory. The monitoring of the patient's comorbidities is necessary. Close observation for possible compartment syndrome must be undertaken and early fasciotomy needs to be considered or extended if not previously performed in the operating room. This can easily be done under sterile conditions and local anesthesia at the bedside.

Complications

There is an increased risk of mortality related to the multiple comorbidities that are usually present in these patients. Loss of the extremity is a real possibility. Local wound problems are very common.

Follow-Up

These patients require very close follow-up because the potential for rethrombosis is high. They need to be followed frequently during the first year postoperatively. After discharge they may need to be followed at least weekly depending on their problems. These patients will have more wound problems and their comorbidities can be more of a problem. This requires close consultation with the primary care physician. Frequent physical exams and duplex scanning will be necessary. The patient's condition will dictate the frequency. After the first year they should be able to be followed more routinely.

Chapter 106

Fasciotomy: Lower Extremity

Mark K. Eskandari

Indications

Indications for a fasciotomy of the lower extremity include acute compartment syndrome after significant lower extremity trauma, prolonged ischemia from acute occlusion, or excessive trauma related to strenuous exercise.

Preop

Most of these procedures are performed under general anesthesia. However, they may be performed under spinal or epidural anesthesia with the patient in a supine position. Additional preoperative assessment includes the use of compartment pressures. Anything above 40 mm Hg is diagnostic of acute compartment syndrome. Typically, the diagnosis is made on clinical grounds alone with evidence of swelling over the compartments, the anterior being the most predominantly affected, followed by the lateral, deep posterior, and superficial posterior. Clinically, the patient will have exquisite tenderness with palpation of the particular compartment and pain on passive plantar and dorsiflexion with normal vascular examination and normal overlying skin and sensory examination.

Procedure

Step 1. The patient is in a supine position. Most fasciotomies are performed through two incisions: one incision is medial and one incision lateral. The lateral incision is approximately 5-7 cm in length, about 2 cm posterior to the posterior aspect of the tibia carried down through the subcutaneous tissue.

Step 2. Both the anterior and the lateral fascial compartments and the septum separating these two are identified. The anterior compartment is then incised using electrocautery, and the whole length of the compartment is decompressed using scissors incising the fascial band, both proximally and distally.

Step 3. A similar technique is utilized for decompression of the lateral compartment, taking care to limit the amount of fascial incising proximally because of the location of the superficial peroneal nerve just beyond the head of the fibula on the lateral aspect of the leg.

Step 4. The posterior compartments, the deep and superficial, are approached through a medial incision. The incision is approximately 5-7 cm in length and approximately 2 cm beyond the posterior aspect of the tibia in the soft part of the leg. Through this longitudinal incision, frequently the saphenous vein will be identified and care should be taken to preserve this if possible.

Step 5. The superficial fascia is incised, retracting the gastrocnemius muscle posteriorly.

Step 6. The deep compartment is then decompressed by taking down the attachments of the soleus muscle to the posterior aspect of the tibia using electrocautery. This should be done carefully because of the proximity of the posterior tibial and peroneal vessels just below the fascial band. This is incised to the length of approximately 5 cm.

Step 7. Hemostasis is obtained and most fasciotomy sites are packed open with a moist gauze.

Step 8. After approximately 3 days, the patient may be returned to the operating room for primary closure, split-thickness skin graft, or these wounds may be allowed to close secondarily.

Postop

Additional postoperative care includes identifying postoperatively the integrity of the superficial peroneal nerve and the ability to dorsiflex the foot. Many of these patients with prolonged periods of ischemia and acute compartment syndrome can develop rhabdomyolysis so renal function and CPK levels should be assessed and treated appropriately.

Complications

Complications related to four-compartment fasciotomies include, first and foremost, injury to the superficial peroneal nerve during the decompression of the lateral compartment; wound infection; and bleeding from the site, particularly in patients who were anticoagulated, necessitating adequate hemostasis at the completion of the procedure.

Chapter 107

Toe Amputation

Mark K. Eskandari

Indications

Toe amputations are generally indicated for wet gangrene, dry gangrene, and/or osteomyelitis of the toe.

Preop

The procedure is usually performed in the operating room, with local anesthesia or an ankle block. While in a supine position, the foot to the knee is prepped and draped in a sterile fashion. Systemic intravenous antibiotics are administered immediately prior to the procedure.

Procedure

Step 1. The local anesthetic is given if desired using 1% lidocaine without epinephrine on either side of the web spaces of the desired toe amputation site. Or, if an ankle block or a spinal has been administered, this should be tested.

Step 2. If the toe amputation is to be either the second, third, or fourth toe, an elliptical incision is made with the apices of the incision on the dorsum and plantar aspect of the foot and the toe amputated either at the proximal phalanx site or proximal to the metatarsal head.

Step 3. The incision is carried down to the level of the bony tissue. Hemostasis is obtained at the digital vessels and the bone transected using a bone cutter. If the level of amputation requires resection of the metatarsal head, the joint space is entered and the metatarsal head resected back using rongeurs.

Step 4. The ligaments and tendons are transected back as far proximally as possible.

Step 5. The wound is irrigated with saline solution. If it is dry gangrene or a clean wound, it may be closed primarily with interrupted vertical mattress 3-0 nylon sutures. If it is infected or a wet gangrene amputation site, it is packed open with a moist saline-soaked gauze.

Step 6. If the toe amputation is of either the first or second toe, a racquet-shaped incision is utilized as opposed to an elliptical incision and the amputation performed as described above.

Postop

Postoperatively, these patients should be evaluated to determine the causative factors of their gangrenous toes.

Northwestern Handbook of Surgical Procedures, edited by Richard H. Bell, Jr. and Dixon B. Kaufman. ©2005 Landes Bioscience.

Complications

Complications from this procedure are early toe amputation without revascularization leading to proximal skin necrosis. Therefore, an adequate assessment of perfusion to the foot should be delineated with arterial blood flows prior to any toe amputation unless the foot is septic.

Follow-Up

After wound healing is complete, patients should be referred to a prosthetist for shoe-wear evaluation. Patients should be given appropriate deep venous thrombosis prophylaxis until they are mobile.

Chapter 108

Transmetatarsal Amputation

Nancy Schindler

Indications

Most commonly performed for gangrene or nonhealing ulcer of the toes. If amputation of more than two toes is required and gangrene extends proximal to the metatarsophalangeal joint, transmetatarsal amputation is indicated. Lesions may be due to emboli, atherosclerosis, or neuropathy.

Preop

Most important is the physical exam and noninvasive testing to determine if the patient has adequate blood flow to heal the wound. A variety of tests are available to assess this, including ankle-brachial index, arterial Doppler, pulse volume recordings, transcutaneous oximetry, and laser Doppler. Diabetic patients need attention to glucose control perioperatively. Anesthesia may be regional (spinal or ankle block) or general.

Procedure

Step 1. Note: Gentle tissue handling throughout the operation is essential for successful wound healing.

Step 2. An incision is made on the dorsal surface of the foot at the midmetatarsal level.

Step 3. Extend the incision to create a plantar flap, which should be as long as possible while preserving only viable tissue.

Step 4. The dorsal incision is carried down to the bone, dividing the extensor tendon as proximally as possible.

Step 5. Metatarsal bones are divided with a saw, just proximal to the plantar foot incision.

Step 6. Remaining tendons are pulled taut and divided as proximally as possible with a scalpel.

Step 7. The plantar flap is then cut to the appropriate size so that it can be rotated anteriorly without tension.

Step 8. The subcutaneous tissue is approximated with interrupted absorbable suture.

Step 9. The skin is closed with interrupted monofilament suture, being careful not to tie the sutures too tightly.

Step 10. A bulky protective dressing is applied.

Northwestern Handbook of Surgical Procedures, edited by Richard H. Bell, Jr. and Dixon B. Kaufman. ©2005 Landes Bioscience.

Postop

Patients should keep the foot elevated and should not bear weight on the foot for 10-14 days. Patients should be given appropriate deep venous thrombosis (DVT) prophylaxis until they are mobile.

Complications

Failure to heal, wound infection, DVT, pulmonary embolism.

Follow-Up

After wound healing is complete, patients should be referred to a prosthetist for shoe-wear evaluation. Patients should be given appropriate DVT prophylaxis until they are mobile.

Chapter 109

Below Knee Amputation (BKA)

Nancy Schindler

Indications

Gangrene or nonhealing foot wounds beyond the mid-foot. Rest pain or ulcer not amenable to revascularization due to anatomy, conduit availability, or patient condition comprise the indications for below knee amputation (BKA). Below knee amputation is contraindicated in debilitated, bedridden patients with knee contractures.

Preop

Most important is the physical exam and noninvasive testing to determine if the patient has adequate blood flow to heal the wound. A variety of tests are available to assess this including ankle-brachial index, arterial Doppler, pulse volume recordings, transcutaneous oximetry, and laser Doppler. Diabetic patients need attention to glucose control perioperatively. Anesthesia may be regional or general.

Procedure

Step 1. Note: Gentle tissue handling throughout the operation is essential to successful wound healing. The procedure may be performed with a tourniquet if desired.

Step 2. The entire leg is prepped and draped. If possible any open or infected wounds should be covered with an occlusive dressing.

Step 3. The anterior incision is made approximately 10 cm below the tibial tuberosity.

Step 4. A long posterior flap incision is made.

Step 5. The anterior incision is carried down to the tibia. The anterior vessels are identified in the lateral wound. The artery and vein are identified, ligated, and divided. The nerve is placed under tension and divided.

Step 6. The musculature of the leg is divided with cautery and the fibula is exposed. It must be exposed about 5 cm above the skin incision.

Step 7. The musculature of the medial leg is divided with cautery. The posterior tibial and peroneal vessels are identified, ligated, and divided.

Step 8. The tibia is then exposed and the periostium is elevated. The tibia is divided transversely 1-2 cm above the anterior skin incision. The fibula is then divided 1-2 cm above the point of transection of the tibia.

Step 9. A small bevel of anterior tibia should be sawed off at a 45° angle.

Step 10. The flail leg is then placed on mild traction. The tibial nerve is placed on mild tension, ligated, divided, and allowed to retract.

Northwestern Handbook of Surgical Procedures, edited by Richard H. Bell, Jr. and Dixon B. Kaufman. ©2005 Landes Bioscience.

Step 11. An amputation knife is then used to create the posterior flap of gastrocnemius and soleus muscles. Long smooth strokes are used and the flap is tapered.

Step 12. A rasp or saw is used to smooth the surface of the tibia.

Step 13. The posterior flap is rotated anteriorly and interupted absorbable suture is used to close the subcutaneous tissue.

Step 14. Interupted monofilament suture or staples are used to close the skin.

Step 15. A dressing is applied that will keep the knee extended. A knee immobilizer or cast can be used for this purpose.

Postop

Physical and occupational therapy consultation should be obtained as soon as possible in the postoperative period. The leg should be kept in some type of dressing which maintains knee extension until most of the postoperative pain has resolved. Failure to do so may result in a contracture. Most patients will require an inpatient rehabilitative stay if they are candidates for ambulation. A prosthesis can be fitted when wound healing is complete.

Complications

Complications include failure to heal, wound infection, deep vein thrombosis, pulmonary embolism. Because these amputations are frequently performed in patients with severe vascular disease and diabetes, cardiac complications and death can also occur.

Follow-Up

Rehabilitation medicine should be consulted to coordinate physical and occupational therapy as well as to prepare the residual limb for prosthesis if appropriate.

Chapter 110

Above Knee Amputation (AKA)

Nancy Schindler

Indications

Above knee amputation is indicated for gangrene or nonhealing ulcer when perfusion makes healing of a below knee wound unlikely and the patient is not a candidate for revascularization. It is indicated for nonhealing wounds or gangrene in a patient with a knee contracture and for severe sepsis due to gangrene and infection where a lesser amputation may be life threatening. It may also be indicated for severe leg trauma. Above knee amputation requires significantly higher energy expenditure for ambulation with a prosthesis and should be performed only when necessary in a patient who has the potential to ambulate.

Preop

Most important is the physical exam and noninvasive testing to determine if the patient has adequate blood flow to heal the wound. A variety of tests are available to assess this including ankle-brachial index, arterial Doppler, pulse volume recordings, transcutaneous oximetry, and laser Doppler. Diabetics need attention to glucose control in the perioperative period. Anesthesia may be regional or general.

Procedure

Step 1. Note: Gentle tissue handling throughout the operation is essential for successful wound healing.

Step 2. A wide fish-mouth or circumferential incision is made in the skin.

Step 3. The greater saphenous vein is identified, ligated, and divided.

Step 4. Muscles are cut using cautery and are allowed to retract upward.

Step 5. In the medial leg, the femoral artery and vein are identified. They are ligated and divided individually. The proximal end is suture ligated.

Step 6. The femur is exposed and a periosteal elevator is used to expose bone about 10 cm above the incision.

Step 7. In the posterior thigh, the sciatic nerve is identified. This is placed on traction and ligated as high as possible. It is then divided and allowed to retract high into the thigh.

Step 8. The femur is divided about 10 cm above the skin incision.

Step 9. A rasp or the saw is used to round off the edges of the bone.

Step 10. The fascia is closed with interrupted absorbable suture.

Step 11. The skin is closed with interrupted monofilament suture or staples.

Step 12. A soft bulky dressing is applied.

Northwestern Handbook of Surgical Procedures, edited by Richard H. Bell, Jr. and Dixon B. Kaufman. ©2005 Landes Bioscience.

Postop

Physical and occupational therapy consultation should be obtained as soon as possible in the postoperative period. Most patients will require an inpatient rehabilitative stay if they are candidates for ambulation. A prosthesis can be fitted when wound healing is complete.

Complications

Failure to heal, wound infection, deep venous thrombosis, pulmonary embolism. Because these amputations are frequently performed in patients with severe vascular disease and diabetes, cardiac complications and death can also occur.

Follow-Up

Rehabilitation medicine should be consulted to coordinate physical and occupational therapy as well as to prepare the residual limb for prosthesis if appropriate.

Chapter 111

Varicose Veins

William H. Pearce

Indications

Varicose veins are a common clinical problem. Primary varicose veins occur as a result of saphenofemoral or saphenopopliteal valvular incompetence. Surgery is indicated for cosmesis, pain, swelling, stasis dermatitis, bleeding, and venous ulceration.

Operative Principles

Obliteration of significant sources of venous superficial reflux is the basis for this operation. It is important to identify the location of significant reflux. Reflux may arise at the saphenofemoral or saphenopopliteal or in intermediate locations where perforating branches join the superficial system. Once the reflux has been identified and ligated, enlarged tributaries are avulsed or stripped.

Preop

All patients undergoing excision of varicose veins should undergo duplex scanning. Duplex ultrasound scanning serves two functions. First it ensures that the deep residual veins are competent and normal. Patients with significant deep venous disease should not undergo stripping unless under special circumstances. Secondary duplex scans should be used to identify significant sources of reflux. The entire course of the greater saphenous and lesser saphenous veins is followed. Valsalva maneuvers are performed to identify areas of reflux; these areas may be marked preoperatively by the blood flow technician. When the patient arrives for the operative procedure, the veins are marked with a water-insoluble marker to identify all perforators and dilated venous tributaries.

Procedure

Step 1. For patients undergoing greater saphenous stripping, the patient is placed in the supine position. The patients undergoing lesser saphenous stripping are placed face down.

Step 2. The patients are prepped using a benzalkonium chloride 1:750 (Zephiran™) solution. This clear solution allows the previous markings to be identified.

Step 3. A small cutdown is performed on the source of reflux. In the saphenofemoral region, a small, transverse incision should be made just medial to the femoral artery just below the groin crease. The saphenofemoral junction is identified by tracing the saphenous vein to its confluence with the common femoral vein. All of the tributaries are divided (superficial inferior epigastric, superficial external pudendal, superficial circumflex iliac, and arteriolateral superficial veins).

Northwestern Handbook of Surgical Procedures, edited by Richard H. Bell, Jr. and Dixon B. Kaufman. ©2005 Landes Bioscience.

Step 4. Once the common femoral vein has been identified, it is safe to ligate the saphenous vein close to the common femoral vein. An external stripper may be placed either antegrade or retrograde through the saphenous system. The stripping system is tied to the saphenous vein, generally in the groin. If the catheter is placed retrograde a separate stab incision is made just above the knee. The greater saphenous is stripped from groin to knee using an inversion technique.

Step 5. Occasionally, the greater saphenous may be completely stripped from groin to ankle. It is not often recommended to strip the greater below the knee since injury to the saphenous nerve is more common.

Step 6. Clusters of varicose veins are removed using the stab avulsion technique. Using an 11 blade a small incision less than 1/8 inch is made directly over the vein. Either using a vein hook, or a small hemostat, the vein is grasped and teased from the underlying tissue until the vein is avulsed. Generally, these small stab incisions need only to be closed with a steri-strip or a small 5-0 subcuticular suture. The leg is milked of any residual hematoma and the groin wound is closed in multiple layers with the final layer consisting of the 5-0 subcuticular absorbable suture.

Step 7. Should the patient require lesser saphenous stripping, the patient is placed prone on the operating room table. An incision is made directly over the saphenopopliteal junction. This junction should be identified preoperatively with duplex ultrasound. For the stripping of the lesser saphenous it is important to realize that injury can occur to either the superficial peroneal or sural nerves. Therefore it is very important to clearly see the lesser saphenous vein before stripping. Stab avulsions are made in the lower segments of the lesser saphenous vein.

Step 8. Following completion of the procedure, the wounds are steri-stripped and a dressing of Kerlix and an external Ace wrap is placed from the metatarsal heads to the mid thigh.

Addsteps. To avoid injury to structures other than the vein to be stripped, meticulous dissection and clear identification of the structures is mandatory. Since inadvertent stripping of arterial and other venous structures has occurred, it is mandatory to identify the saphenofemoral junction and see the common femoral vein. A clear knowledge of the anatomy is important.

Postop

Postoperatively, the patient is discharged to home. The Ace wrap is kept in place for 24 hours. After 24 hours the Ace wrap is applied from the metatarsal heads to the tibial tuberosity. This wrap is used for approximately 2 weeks until the patient returns to the office.

Complications

The most common complication is bruising throughout the leg. The patient should be warned of this complication since it is so common. Injury to peripheral nerves including the saphenous, sural, and peroneal nerves may occur. The sensory neuropathy that occurs following these injuries will generally resolve over one year. However, motor injuries are usually permanent. Seromas occasionally occur along the course of a saphenous vein but are uncommon. Wound infections are also uncommon (<5%). Deep venous thrombosis is rare (0.6%). Fatal pulmonary embolism has been reported to occur in 1 in 30,000. In addition, older literature has reported inadvertent stripping of arterial structures or deep venous structures.

Follow-Up

The patient is seen 2 weeks following the surgical procedure. The sutures are removed; however, in most instances subcuticular stitches are placed so this is not required. No further follow-up is required. The patients are encouraged to wear support hose since they generally have a predisposition for venous disease.

LANDES BIOSCIENCE

Vademecum

More Handbooks in this Series

How to order:

- at our website www.landesbioscience.com
- by email: orders@landesbioscience.com
- by fax: 1.512.637.6079; by phone: 1.800.736.9948
- by mail: Landes Bioscience, 1002 West Avenue, 2nd floor, Austin, TX 78701

Chemotherapy Regimens and Cancer Care
Alan D. Langerak, Luke P. Driesbach
Video-Assisted Thoracic Surgery
Todd L. Demmy
Cardiothoracic Surgery 3ed
Fritz J. Baumgartner
Hemostasis and Thrombosis 2ed
Thomas G. DeLoughery
Perioperative Care in Cardiac Anesthesia and Surgery
Davy C.H. Cheng, Tirone E. David
Clinical Transfusion Medicine
Joseph Sweeney, Yvonne Rizk
Vascular Disease 2ed
Darwin Eton
Bone Marrow Transplantation
Richard K. Burt, Scott T. Lothian, H. Joachim Deeg, George W. Santos
Organ Transplantation 2ed
Frank P. Stuart, Michael M. Abecassis, Dixon B. Kaufman
Transplantation Drug Manual 4ed
John Pirsch, William Simmons, Hans Sollinger
Organ Procurement and Preservation
Goran B. Klintmalm, Marlon F. Levy
Aortic Surgery
Jeffrey Ballard
Surgical Oncology
David N. Krag
Cardiac Pacemakers and Defibrillators 2ed
Charles J. Love
Breast Diseases
Patrick I. Borgen, Arnold D.K. Hill
Microsurgery
Konstantinos Malizos
Trauma Management
D. Demetriades, Juan A. Asensio
Musculoskeletal Procedures: Therapeutic and Diagnostic
Jacqueline Hodge
GI Endoscopy
Jacques Van Dam, Richard Wong
Ultrasound for Surgeons
Heidi L. Frankel
Nuclear Medicine
William Leslie
Physician's Pocket Lawyer
Charles Ellington, John Bradley
Urologic Oncology
Daniel A. Nachtsheim
Vignettes on Surgery, History and Humanities
Luis Toledo-Pereyra
Whose Disease? Eponymous Pathology
Hugh McLaughlin

Access for Dialysis: Surgical and Radiologic Procedures, 2ed
Ingemar J.A. Davidson
Dermatopathology
Ramon Sanchez, Sharon Raimer
Extracorporeal Life Support
Dan M. Meyer
Endosurgery for Cancer
Steve Eubanks, Ricardo Cohen, Riad Younes, Fredrick Brody
Kidney-Pancreas Transplantation 2ed
Ingemar J.A. Davidson
Radioguided Surgery
Eric D. Whitman, Douglas Reintgen
Clinical Immunology
Reginald Gorczynski, Jacqueline Stanley
Mechanical Circulatory Support
Wayne E. Richenbacher
Tropical Dermatology
Roberto Arenas, Roberto Estrada
Pediatric Surgery
Robert Arensman, Daniel Bambini, P. Stephen Almond
Pediatric Laparoscopy
Thom E Lobe
Endocrine Surgery
Richard A. Prinz, Edgar D. Staren
Surgical Arithmetic
Lawrence Rosenberg, Lawrence Joseph, Alan Barkin
Burn Care
Steven Wolf, David Herndon
Hepatobiliary Surgery
Leslie H. Blumgart, Ronald Chamberlain
Liver Transplantation
Michael R. Lucey, James Neuberger
Tropical Neurology
U.K. Misra, J. Kalita, R.A. Shakir
Vital Signs and Resuscitation
Joseph P. Stewart
The Biology and Practice of Current Nutritional Support, 2ed
Rifat Latifi, Stanley Dudrick
Parasites of Medical Importance
Anthony J. Nappi, Emily Vass
Northwestern Handbook of Surgical Procedures
Richard Bell, Dixon B. Kaufman
Pediatric Neurosurgery
David Frim and Nalin Gupta
Emergency Medicine
Sean O. Henderson
Origins of the Knife
Early Encounters with the History of Surgery
Luis Toledo-Pereyra